THE GRUNDIG BOOK

D1440365

A FOCAL SOUNDBOOK

First printed August 1958
Reprinted September 1958
Reprinted November 1958
Reprinted April 1959
Revised Edition October 1959
Reprinted April 1960
Revised Edition April 1961
Revised Edition January 1962
Revised Edition April 1963
Revised Edition August 1964
Revised and Enlarged Edition November 1966
Revised and Enlarged Edition September 1968

Printed in Great Britain, 1968, for
Focal Press Ltd., 31, Fitzroy Square, London, W.1,
by W. & G. Baird Ltd., Belfast, Northern Ireland.

The Grundig Book

FREDERICK PURVES

THE FOCAL PRESS
London and New York

Contents

7

Grundig is the greatest name in the world of popular tape recording and the total sales of Grundig tape recorders can now be reckoned in millions.

Grundig machines have been largely responsible for the growth of popular sound recording. Their success stems from the fact that they make it possible for non-technical people—musicians, business men, housewives, doctors, actors, comedians and clergymen—to make the miracle work for them at the touch of a button without knowing a thing about how it all happens.

The Grundig Book was the first independent publication ever to be written about a single make of tape recorder and it has been more successful than any book of its kind. It tells you in plain language how to exploit all the wonderful possibilities of your Grundig to the full.

At the same time, the one man in a hundred who wants technicalities will find enough in both the main text and the data section at the end of the book to keep him well-informed and happy.

For clarity, diagrams in the main text of this book have been simplified and do not necessarily represent actual designs.

CHAPTER 1

How your Grundig Works

IF YOU DON'T KNOW HOW A TAPE RECORDER works, what the various controls do, and what their names are, this chapter and the next will tell you. It will make it easy for you to understand the chapters that follow. But if you know all this, you can start right away at chapter 3.

The sounds you want to record on your tape recorder are just a form of energy. They are made up of waves in the air like the waves set up when you drop a stone into a pool. You can change these sound waves into electrical or magnetic waves just as you can change money into pennies, cents or pesos and then turn it back again into the coinage you started with.

Storing sound

When you speak into the microphone of a tape recorder it changes the sound energy of your voice into electrical energy. In this form, it can now pass down the microphone cable and into the tape recorder. Inside the tape recorder, the electrical energy is changed into magnetism. So you have now changed your original sound into magnetism which acts on the recording tape and leaves a permanent record, in magnetism, of the sound you started with. In this form, the sound can be stored on the magnetic recording tape until you want to use it again. When that time comes your tape recorder goes into action in reverse and changes the magnetic energy back into electrical energy, feeds it into a loudspeaker and you get back your original sound.

Writing with rust!

So far you haven't heard anything about the tape itself,

and it's time you did because the whole magic of the modern tape recorder lies in the tape.

The tape you use in your Grundig today is amazing stuff. It is a plastic ribbon $\frac{1}{4}$ in. wide and coated with powdered iron oxide—better known as rust.

These tiny particles of oxide preserve the record you make on the tape. Before you make a record all these tiny particles of iron oxide are lying around higgledy-piggledy like a lot of pieces of a jig-saw puzzle; every one is capable of being turned into a magnet—just like pins when you bring them in contact with a real magnet. This is exactly what happens when the tape passes over the sound head. The magnetism created by the sound head turns the oxide particles into

SOUND ON TO TAPE. *Sound waves strike the microphone and change into electricity. The electrical variations are then fed to the sound head in the tape recorder. The sound head changes the electricity into magnetism, which leaves an invisible 'pattern' on the tape.*

magnets and forms their magnetic fields into a pattern—just like sorting out the scrambled bits of the jig-saw puzzle.

You can't *see* the pattern, but it's there just the same, and it changes as you go along the tape according to the amount of magnetism coming from the head. And if you could trace the magnetism back, you would find that it changes according to the way the original sound waves beat upon the microphone. So the pattern you get on the tape represents the sound vibrations that began it all. In other words, you have made a record of the original sound.

When you want to play back the record, you have only to run the tape once more over the same sound head, and the magic works backwards. As the magnetic pattern on the tape passes over the sound head you get electrical vibrations in the sound head. They follow the same pattern as the electrical vibrations that you put in at first. All you have to do is to feed that electrical current into the tape recorder speaker and you get your sound back again.

Finally, one of the amazing things about your Grundig is that when you are tired of a particular recording you can 'wipe' it off and start all over again. This is done by another type of head called an erase head. The erase head simply shakes up all the bits of the jig-saw—the magnetic pattern— and destroys the picture. It is fed with a high frequency electric current that creates a rapidly changing magnetic field. As the tape passes over it, all the tiny molecules in the magnetic oxide get thoroughly scrambled and have all their magnetism taken away. So you are left with a 'blank' tape

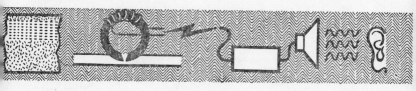

SOUND FROM TAPE. The magnetic pattern passes over the sound head and sets up an electric current in the winding of the head. The electric current is magnified in the amplifier and fed into the speaker, which changes it back into sound waves similar to the ones that made the record.

all ready for making a record. The whole thing is automatic so that as soon as you start another recording, the erase head goes into action and wipes the tape clean before it reaches the recording head.

Well now that you know how the magic works you will be ready to take a closer look at the magic box itself.

The main units of your Grundig

The two principal parts of your Grundig are the tape deck and the amplifying unit. The tape deck looks after the purely mechanical business of handling the tape; the amplifying unit does all the electronic side of the recording and playback.

11

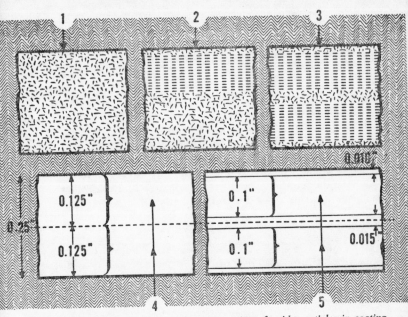

THE TAPE. 1 Diagrammatic impression of oxide particles in coating before recording. 2 After recording track 1. 3 After recording track 2. 4 and 5 Standard dimensions of magnetic recording tape and arrangement of the double track.

Everything that helps to move the tape past the sound heads when you record or play back, and any control knobs, levers or buttons that you use in the process belong to the tape deck.

The amplifying unit does all the electrical work. When you make a record it amplifies the signal you want to record and feeds it into the recording head at the right strength to give first class quality. When you play back the tape it takes the weak signal that comes from the playback head and amplifies it so that it comes to you from the loudspeaker as nearly as possible like the original sound you recorded.

The Tape Deck

When you open the lid of your Grundig, you uncover the tape deck. On this you will see two spool holders which may

12

or may not be loaded with tape spools and tape. (If you are unpacking a new machine the full and empty spools will be packed in separate boxes in the transit case). On the latest battery portable model the tape and both spools are housed in a single cassette.

Between and just in front of the spool holders you will see a cover with a slit running along its length. This is the sound channel cover and in front of it on the current models you will find the tape controls—Stop, Start, Fast Forward and Fast Reverse—set out like a row of piano keys with the magic eye recording level indicator in the middle. On earlier models the controls took other forms and were laid out differently. And on one of the mains/battery portables the deck stands upright and the controls are set out along the top. However, no matter whether they are knobs, keys, levers or buttons, or whether you push, turn or press them, the controls do the same job on all models. These jobs are dealt with one by one below—

START KEY: The start key starts the tape moving and it switches on either the recording or playback circuit. It starts the tape by closing up the capstan and pressure roller so that they grip the tape and draw it through the sound channel. At the same time, it connects the take-up spool to the motor through a slipping clutch. This clutch is adjusted so that it slips as soon as the tape is reasonably tight. So, as the capstan pulls the tape off the supply spool, the take-up spool winds it up with just the right amount of tension to keep it tidy.

PAUSE KEY: This is the key you press when you are recording or playing back and want to halt the tape for a few moments without stopping the motor or interfering with any of the other controls—e.g. if you are dictating and want to pause while you think of the next word.

When you press the Pause Key the first time it separates capstan and pressure roller, takes the pull off the tape and brings it to rest. It does not affect the pads, which keep the tape in contact with the sound heads. When you press

13

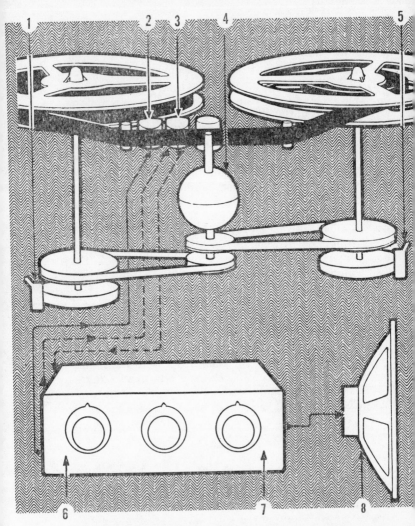

SIMPLIFIED TAPE RECORDER DIAGRAM. 1 Supply spool brake. 2 Erase head. 3 Record playback head. 4 Motor, capstan and pressure roller assembly. 5 Take up spool brake. 6 Input to amplifier. 7 Output from amplifier to speaker. 8 Speaker. Clutches in spool pulleys engage drive as required.

it the second time it closes up the capstan and pressure roller, grips the tape and starts it moving once more. The first pressure leaves the key down; the second pressure returns it to the raised position. So remember—if you leave this key down, you can't record or play back.

STOP KEY: This is the key you press when you are finished recording or playing back for the time being. It opens up the capstan and pressure roller and leaves them apart so that there is no pull on the tape even when the motor is switched on and the capstan is turning—as it is on most, but not all, Grundig models. Pressing this key returns all other keys to the off position. It also raises the pressure pads that keep the tape in contact with the record/playback head so that you can remove the tape or fit a new one. And finally, it breaks the electrical contact between the record/playback head and the amplifier.

Pressing the Stop Key brings the tape to rest whether you are recording, playing back or winding the tape fast in either direction, and it applies a brake to the spools so that they come to rest quickly.

You should always make sure that you have pressed the Stop Key before you put your machine away. If you forget, you leave the rubber pressure roller pressed against the capstan and by the next time you use your machine there will be a flat spot impressed in the surface This will produce a wavering effect on steady notes and it may be some time before it irons itself out.

RECORD SELECTOR: This selector may be a press key in the bank of keys along the front of the deck as in most of the current models or it may be a button on top of the deck. Pressing this key connects the recording circuit to the record/playback head. You press it down when you want to make a record, just before you press the start key to start the tape moving. On some models you have to keep the record key held down until you have pressed the start key. It will then remain locked automatically in the down position keeping the circuit switched to record. On other models, the first

15

pressure on the key locks it down and connects up the recording circuit, and the next pressure releases it. But the controls are always interlocked so that it is impossible to operate the recording switch while the tape is running. You can only engage it when the tape is at rest.

The record key is automatically returned to the off position every time you press the stop key. You have to press it again if you want to go on recording. On some models that have a press button instead of a key, you turn the button to lock it in the on position.

RECORD SAFETY BUTTON: Models which use a separate control for selecting record or playback have a red press button that you must press down before you can turn the control into the record position. This is a safety measure to stop you from accidentally erasing the tape.

PLAYBACK SELECTOR: On most models, to play back the tape you just press the Start key or turn a multi-position control to the Start position, depending on your model. When you do this the tape starts moving and the circuit is automatically switched to play back the tape. On some early models there is a key to select the playback function and a separate key to start the tape.

FAST REVERSE: This is the control you use when you want to wind back the tape you have just recorded or played back. It brings into operation a specially powerful and fast drive to let you cut the time of rewinding as short as possible.

On some models the Fast Reverse is controlled by a key which stays down when you press it, so that rewinding is an all-or-nothing operation. Other models have a slider which take up the drive gradually and lets you 'inch' the tape.

FAST FORWARD: This control is a press key, lever or slider as for the Fast Reverse, but it winds the tape forward, i.e. from left to right. You use it when you want to skip over part of the tape and start recording or playing back at a point

16

further on. Once you have operated the control, you can only cancel it by pressing the normal stop key. When you do this the drive is disengaged and the brakes are applied, bringing the spools to a standstill almost instantly and without forming slack loops of tape.

COMBINED FASTWIND AND INCHING CONTROL: While some models have separate press keys for Reverse and Forward Fastwinding, others have a slider which you move to the left to take the tape back and to the right to take it forward. This type of control takes up the drive gradually so that you can 'inch' the tape a short distance to right or left by small movements of the slider. This is useful for pinpointing your position accurately on the tape. When the slider is fully engaged to one side or the other, pressing the stop kep returns it to neutral and stops the tape.

SPEED CONTROL: According to the model of your Grundig, it will record and play back at one or more of the following tape speeds (measured in inches per second—i.p.s. for short): $1\frac{7}{8}$, $3\frac{3}{4}$, $7\frac{1}{2}$. Some models have only one recording speed—$3\frac{3}{4}$ i.p.s., others have two, either $1\frac{7}{8}$ and $3\frac{3}{4}$ or $3\frac{3}{4}$ and $7\frac{1}{2}$, while others again will record at all these speeds.

Where there is a choice of speeds, you have a control on the deck which lets you select one or the other. The type of control depends on the model: most current multi-speed machines use a thumbwheel, which incorporates the on-off switch and has the speeds marked on the edge.

When you record a tape at a particular speed, you must play it back at that speed or it will sound all wrong. For instance, if you record at $3\frac{3}{4}$ i.p.s. and play back at $7\frac{1}{2}$ i.p.s., the pitch of the sound will jump an octave as well as running at twice the speed.

Broadly speaking, the faster speeds give the best quality. but they use up your tapes quicker. For 'Pop' quality recording and speech, $1\frac{7}{8}$ and $3\frac{3}{4}$ are all you need. For the finest quality orchestral records, it pays to run at $7\frac{1}{2}$ i.p.s. and put up with the higher tape consumption.

The Sound Channel

Between the spool holders you will see a cover with a slit running along it. This is the cover over the magnetic heads that record, play back and erase the tape; it also covers the capstan and pressure rollers that pull the tape over the heads. This complete arrangement is known as the Sound Channel.

On some models the cover is part of the top of the deck and you can't see under it without removing the whole of the deck. This job is best left to your Grundig dealer if it ever needs to be done. On most of the current models, however, the sound channel cover is simply clipped or sprung into position, and you only need to squeeze the ends together and you can ease the cover off. This is useful when you want to clean the heads (p. 157) or if the tape sticks and snarls up (p. 160). Now for a brief description of each of the components in the Sound Channel:

CAPSTAN ASSEMBLY: This consists of two rollers that grip the tape and pull it through the sound channel in the same

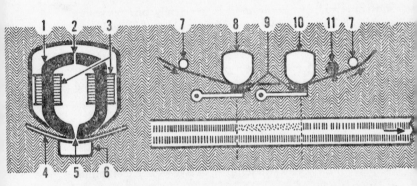

SOUND CHANNEL. Record/Playback Head (left) *comprises: 1 Magnetic core; 2 back gap; 3 head coil; 4 tape; 5 front gap; 6 pressure pad to hold tape in contact with front gap.* Sound Channel (right top) *comprises: 7 Tape guide pillars; 8 erase head; 9 pressure pads; 10 record/ playback head; 11 capstan and pressure wheel. Tape below showing lower track recorded; upper track carrying old recording is erased at 8 and new recording added at 10.*

way as the wringer of a washing machine pulls a length of cloth between its rubber faced rollers. One of the rollers—the capstan—is formed by the top end of a steel spindle driven by the capstan motor. The other is a rubber roller that is normally held away from the capstan spindle to let you thread the tape between and allow it to run through freely when you are fastwinding it from one spool to the other.

When you switch on the tape recorder motor, the capstan spindle starts to turn, but it has no effect on the tape until you press the start key. This moves the rubber roller into contact with the capstan spindle, squeezing the tape between the two and drawing it through at a steady speed. As it does this, a gentle braking action is applied to the tape supply spool on the left to stop it from spinning freely and throwing off loose turns of tape. On the other side of the capstan, the take-up spool winds up the tape as it comes out of the sound channel. The take-up spool is driven through a slipping clutch which is always trying to turn faster than the tape will let it. When the pull on the tape reaches a pre-set figure, the clutch slips and the tension in the tape stays constant from beginning to end, irrespective of the diameter of the tape load on the spool.

ERASE HEAD: The erase head is the first head the tape passes over as it runs through the sound channel. In principle, it consists of a C-shaped core of metal with the free ends brought close together, but leaving a narrow gap. A coil of insulated wire is wound around the body of the C. The tape runs across the gap and is held in close contact with it by a spring loaded pressure pad (below).

When you are recording, a high frequency current generated by a valve oscillator is passed around the coil. This creates an intense alternating magnetic field, which completely erases any signal on the tape as it passes the gap. When you switch to play back, no current flows in the erase head, so the record on the tape is not affected.

If you have a Trick or Superimpose switch (p. 28) on your model, it lets you disconnect the erase head so that you can make a second record on the tape without automatically erasing the first.

RECORD/PLAY HEAD: This is the real heart of your Grundig. It can put a record on the magnetic coating of the tape, and it can pick it up and play it back through the amplifier and speaker. It is the same type of head as the one that erases the tape—a C-shaped metal core wound with a coil of insulated wire—but in this case the gap between the open ends of the C is very much finer—no more than two or three ten thousands of an inch. It is important for the gap to be narrow because its width decides how far you can go up the frequency scale; the finer the gap the higher the frequencies the head will record on the tape.

When you make a record, the signal from your microphone pickup, or radio, is amplified and fed into the head coil which turns it into magnetic fluctuations in the gap, just where the tape runs over it. These fluctuations make a record on the magnetic coating of the tape.

When you play back the tape, the head coil this time is connected through the playback amplifier to the speaker. As the tape carrying the magnetic record runs over the gap in the head, the fluctuations of the magnetism in the tape add to or subtract from the magnetic field in the gap. This rise or fall of the magnetism in the head gives rise to electrical fluctuations in the head coil. These voltage fluctuations are fed first into the amplifier, which gives them a boost and then into the speaker, which turns them into sound.

PRESSURE PADS: The tape is held in contact with the front face of the erase and record/play heads by a spring loaded pressure pad—or on some models by a sling of plastic material which wraps around the head and applies an even pressure on each side of the gap. This gives extremely close contact—specially important on 4-track machines (p. 34). On these machines the record on the tape is only ·043 inches wide, so any break in contact between the head and the tape has a very noticeable effect on the reproduction quality.

When you are recording or playing back, the pressure pads hold the tape firmly against the heads. When you press

20

the stop key or engage the fast reverse or forward wind, the pressure pads are raised, leaving the tape free. When you press the pause key, the pads stay down so that there is no gap between the points where the tape stops and starts again.

TAPE GUIDES: At each end of the sound channel, there are metal or plastic posts which guide the tape as it comes off the spool, so that it always enters and leaves the sound channel at the same angle, no matter how little or how much tape there is on the spool. On some models one of the tape guides forms a contact for the automatic stop mechanism (below).

THE TAPE POSITION INDICATOR: All Grundig models with the exception of some earlier types have indicators which show how much tape you have used during recording or playback.

On all current models the indicator is of the digital type in which three figures behind a window on the deck register revolutions of the supply spool up to 999. On some early models the indicator is of the clock face type.

The principal job of the tape position indicator is to let you find any particular place on the tape when you want it. You have to remember to set the indicator back to zero at the start of the spool. After that you only need to jot down the reading of the indicator at any time from the start to the finish of the tape to be able to find that place again. You may, for instance, have a number of short items recorded on one tape; the indicator enables you to select any one of them to play back so long as you have noted down the readings at the start of each item.

You can also use your tape indicator to tell you how long you have been recording and how many minutes there are left before you come to the end of the tape. (This is helpful when you want to record an item on a part-used tape— e.g., a broadcast item where you know the exact length of time it will take.) The chart in the handbook that comes with your machine shows the running time corresponding to the indicator reading.

Old models which have no mechanical counter or dial have a scale marked on the deck between the centres of the spools. You read the scale directly below the edge of the tape build-up on the spool; this gives you a figure which is approximately the number of minutes the tape has been running since the start (assuming that you started with a full spool) and the number of minutes running time left on a full-size spool. The running time refers to standard tape and you have to increase it by half if you are using Grundig long-play tape, double it for double play tape and so on.

TAPE CLEANER: A $\frac{1}{4}$-track tape record is only ·043 wide, and specks of dust on the tape can produce sudden gaps in the signal as they pass over the head. So you need to keep the tape clean if you want first-class reproduction. This is made easy for you on many of the 4-track models by providing a tape cleaner consisting basically of felt pads which can be pressed lightly against the tape. On one model there is a plastic fork fitted with two felt rings which is plugged into holes provided in the top of the deck so that the tape runs between the felt surfaces. On other models, there is a spring loaded, felt-faced roller under the sound channel cover. The roller can be released and brought into contact with the tape by pressing a button on top of the deck. To clean the tape you simply fit or release the cleaner and run the tape over it by operating the fastwind control. Once cleaned the tape should be carefully protected from dust by keeping the spool in a container when not in use.

Tape Transport

The mechanical department of your Grundig has to take care of the tape under three different sets of conditions—playing or recording, winding back at speed and winding forward at speed. It also has to bring the spools to a standstill when you press the stop key and do it without forming loops of slack tape—or breaking the tape.

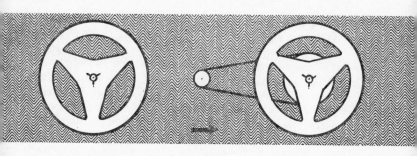

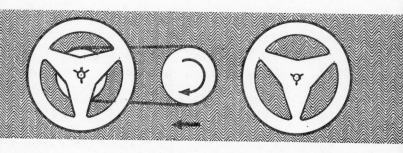

TAPE TRANSPORT SYSTEM. Top: When you record or play back the take-up spool is driven by a clutch which slips as soon as the tape goes taut. Centre: When you wind the tape back the supply spool is driven in reverse at high speed to save time. Bottom: When you wind the tape on, the same high speed drive is employed to get the job done quickly.

WHEN YOU PLAY BACK THE TAPE or make a record on it, this is what goes on: The supply spool on the left is checked by a lightly applied brake which allows it to turn as the capstan draws the tape off it, over the heads and through the sound channel. The take-up spool on the right winds up the tape as it comes out of the sound channel. It is driven by a plastic belt from the capstan motor. The tape spool is always trying to take up the tape faster than it comes out of the sound channel, but there is a safety clutch in the drive which starts to slip as soon as the tape goes taut. So the tape winds off one spool and on to the other under an even tension and at a speed that is governed by the capstan alone and is not affected by the diameter of the tape on either spool. (This is an essential condition because if the speed of the tape changed, the pitch of the recording would change with it and nothing would sound right.)

WHEN YOU WIND THE TAPE BACK after recording or playing, the capstan and pressure rollers open, the pressure pads lift off the heads, and the tape is left free. At the same time, the spool on the left turns at high speed and pulls the tape off the one on the right against the tension of a very light brake. So the tape winds rapidly back to the beginning without wasting a lot of time.

WHEN YOU WIND THE TAPE FORWARD the same thing happens, but this time it works in reverse and the tape winds rapidly on to the right hand spool.

WHEN YOU PRESS THE STOP KEY or when you reach the end of the tape, and the automatic stop comes into action, the drive is disconnected from the spools and the capstan and at the same time a powerful, quick-acting brake brings the spools to a standstill.

Note: On the cassette model both spools are contained in a plastic case but reverse and forward winding operations are the same as on a normal deck.

The Amplifier

The amplifying section of your Grundig is the part that looks after the signal just as the tape deck looks after the tape, When you are recording, the amplifier takes the signal from the microphone, pickup, radio or whatever the signal source happens to be, and adjusts the strength and other characteristics so that it is just right for making a good, high quality record on the tape. It usually has to boost the signal, but sometimes it has to cut down a signal that it is too strong so that it does not overload the tape and spoil the record.

When you are playing back, the amplifier has to take the extremely weak signal that comes from the playback head, boost it, adjust the proportion of the high and low notes and feed it into the speaker so that you hear a reproduction of the original sound.

On some models there are separate amplifiers for recording and playing back; on others, the same components serve both record and playback amplifiers, but they are connected up differently.

On stereo models (p. 39) there are two amplifiers, one for each channel.

POWER SUPPLY: The power for driving the tape transport motor and providing the current for the amplifier unit circuits comes from the main electricity supply, which must always be A.C. (alternating current) and never D.C. (direct current). The alternating current from the mains is fed into a transformer which (1) supplies alternating current of the correct voltage for operating the motor and (2) provides an independent supply to one or more metal rectifiers. These convert the alternating current into the direct current required by the amplifier and its associated circuits.

Power for the Grundig transistor models (p. 41) comes from their own dry battery pack or from a Mains Pack Adaptor (p. 41).

The controls

It doesn't matter which Grundig model you own, the amplifying department will have a number of control knobs and switches mounted on the deck for adjusting the signal as it passes through the various stages in the amplifying system, and it will have a number of output and other

25

sockets which you can use to connect up to other equipment (p. 29), and it will have one or more input sockets where you can plug in the signals you want to record. The input and output sockets are generally mounted on the same panel on the back of the tape recorder, but some of the earlier models had other arrangements. The various sockets are labelled with standard symbols on the current models; some of the earlier models had the name printed next to the socket.

The amplifier controls that you will find on most Grundig models are described below:

ON-OFF SWITCH: This control connects the tape Motor and the amplifier circuits to the electricity supply so that everything is ready when it is wanted with the motor running and the amplifier valves warmed up. To make things easy for you, the On-Off switch is always incorporated in one of the controls, either the volume control or the speed selector since both these controls must also be in the working position before you can record or play back. You can always tell when you have turned the On-Off switch either on or off, because it makes a clearly audible click.

RECORDING LEVEL CONTROL: This is the adjustable control that comes between the signal you want to record and the record/play head. It is either a knob, a radial arm or a thumbwheel—usually marked with a scale so that once you have found the right setting for a particular subject, you can always go back to it.

The recording level control usually does more than one job. On many models it incorporates the On-Off switch so that you automatically switch the recorder on when you move the control from zero. And in addition to controlling the signal level when you are recording, it may control the volume or the tone when you are playing back.

On most models one recording level control looks after all the inputs, depending on which one is either connected up to a signal source or—if more than one source is connected at the same time—which one you have selected with the input selector switch. However, some models have two recording level controls, one for the microphone input and the other for the radio/pickup input. With these models you can mix two inputs (p. 83).

You use the control to adjust the recording level with the aid of the recording level indicator (below).

RECORDING LEVEL INDICATOR: This is usually called the magic eye because on the earlier models it is round and looks something like an eye, in which the fluctuations of the signal are shown by two luminous segments which close and open as the signal strength rises and falls. On current models the round eye has been replaced by a tubular indicator with a luminous line broken in the middle. The two luminous sections close up as the signal strength rises and part as it falls. In both types the correct signal strength is indicated when the two luminous sections just touch on the strongest signals. If they never come together, the signal strength is too low; if they constantly overlap, it is too high.

AUTOMATIC/MANUAL SWITCH: Automatic models adjust the recording level automatically and both the recording level control and the recording level indicator are absent. The TK23A and TK 23L can be switched for either manual or automatic operation.

VOLUME: This is the control that regulates the sound coming out of the speaker when you play back the tape. It may be a knob, lever or thumbwheel, and it may simply control the volume, or it may incorporate other controls. On many models the volume control is also the recording level control when recording, and the on-off switch.

TONE: On most Grundig models the tone is adjusted by a single knob, lever or thumbwheel on top of the deck. As a rule, turning the control clockwise brings out the treble and turning it anti-clockwise cuts it down. So to give the bass a boost, you turn the tone control *down* and turn the volume *up*. To increase treble, you turn the tone control up.

Some Grundig models have separate controls for the lower, middle and upper frequencies. With these controls

27

you can adjust each of the frequencies independently of the others. These controls have a visual indicator which shows you at a glance how the various frequencies are distributed.

SUPERIMPOSE OR TRICK KEY: This control puts the erase head out of action, so that after you have made one record on the tape you can add another without losing the first—*e.g.*, you can add a musical background to a recorded speech.

MONITOR: The monitor control lets you listen through your Grundig speaker to the sound you are recording. It is a rotary control which usually mutes the speaker completely in the off position. On some models the knob itself works an on-off switch—pulling the knob up switches the speaker off and pushing it down switches it on. The on-off switch operates no matter how far the knob has been turned. This switch is handy when you are recording from the microphone and want to be sure that the sound from the monitor speaker won't feed back into the microphone and start a howl.

The monitor control is usually of the multi-purpose type which does another job when you are playing back, *e.g.* volume or tone control.

MUTING SWITCH: This switch is for switching off the internal speaker when you are playing the tape back through an extension speaker or a high fidelity reproducer. It is usually mounted on the socket panel near the extension speaker output socket. (You needn't switch off the internal speaker, of course, but as a rule you will find it more satisfactory to do this if the extension speaker is in the same room as your tape recorder.)

TRACK SELECTOR: The track selector switch is only needed on the 4-track Grundig (p. 34). On a 2-track machine there is only one head; this records one half of the tape on the first run through and the other half on the second run. On the 4-track models, however, there are two heads, each of which makes two $\frac{1}{4}$-track records. So you start by making two $\frac{1}{4}$-track records with one head and then you switch over to the second head and make two more records. This switch is called the track selector and you use it in the same way whether you are recording or playing back. The track selector usually has a third position in which it connects up both heads so that you can play back two tracks at once after you have recorded them separately. The selector may be an arm with three positions, or the arm may be replaced by two press keys, one for each pair of tracks. With the press key type of selector, pressing down both track keys at the same time plays back both tracks at once.

The automatic stop

The tape decks of most Grundig models incorporate an automatic stop that brings the spools to rest when the end of the tape is reached. The mechanism is operated by the strip of metal foil inserted between the end of the magnetic tape and its leader. This foil is standard on most tapes sold today and on all Grundig tapes (p. 93).

As the metal foil strip runs through the sound channel it completes an electrical circuit between one of the tape guides and, during record or playback, a brass insert on the erase head or, during fast wind, the pillar that holds the tape away from the heads. On some models the tape guide pillar is divided to form two contacts; when the metal foil runs over these it completes the circuit and disconnects the tape drive.

When you load a tape into your machine you must remember to wind it on to the take-up spool until the metal foil insert is clear of the sound channel otherwise as soon as you start the tape the automatic stop will come into action and you will have to make a fresh start.

THE SOCKET PANEL. On this panel you will find all the sockets used for connecting your Grundig to the mains and other electronic equipment.

Most models have three inputs. The location of these varies with the model. The positions and the type of socket are given in the data section at the end of the book. General details are as follows:

MICRO INPUT. This is always the correct socket for the microphone supplied with your machine, but it is not suitable for all microphones. Consult your Grundig dealer before experimenting with a different instrument.

DIODE INPUT. This socket is intended for connecting to a radio diode, a Grundig telephone adaptor or, on some models (see Data Section), a high impedance microphone. It also provides a high impedance output (p. 139).

RADIO L.S. INPUT. This socket is intended for connecting to the extension speaker (low impedance) terminals of a radio set, radiogram, or gramophone record player, or direct to a gramophone pick-up.

Most Grundig models have two output sockets, one for high impedance and the other for low impedance equipment.

LOW IMPEDANCE OUTPUT. This output is intended for connexion to an external loudspeaker or to low impedance earphones On earlier Grundig models, insertion of a jack plug into this socket automatically disconnects the internal loudspeaker. On other models. you can use the muting switch.

HIGH IMPEDANCE OUTPUT. This output connects to an external amplifier or to a Grundig mixer unit. It is also suitable for monitoring with high impedance earphones. The signal at this socket is suitable for feeding straight to an external amplifier without any further correction.

OTHER CONNEXIONS. In addition to the above sockets there are a number of others principally for Grundig accessories. These are:

REMOTE CONTROL. This socket is for a Grundig remote control attachment, which enables the machine to be started or stopped remotely. It takes one of two special types of five-pin plug (according to the model of your Grundig and the type of remote control). Not all Grundig models are provided with this. On some models the remote control can also be made to reverse the tape to repeat the passage just played back.

GRUNDIG DISTRIBUTOR SPEAKER. This particular socket was provided only on one Grundig model to connect the output to the Grundig distributor speaker for imparting extra brilliance to the upper frequencies, particularly when reproducing orchestral music.

EARTH. If your Grundig is one of the models fitted with a two-core cable for connexion to the mains supply, there may also be an earth socket on the panel. This socket is provided to add an earth connexion where one is available. The normal electricity supply earth is suitable. You should never earth your Grundig to a gas pipe.

FUSE HOLDERS. Most models have two screw-in or clip-on fuse holders. These may not be on the socket panel but under the tape deck cover. You can find the correct values from the individual data sheets at the back of the book.

MAINS VOLTAGE ADJUSTER. All Grundig models have a mains voltage adjustment. This may be on the socket panel, but on certain models it is under the tape deck cover. The adjustment is made either by moving a lever or by changing the position of a plug.

MAINS CONNEXION. The two or three-core lead which connects your Grundig to the main electricity supply may be permanently wired to the recorder, or it may be a detachable lead, which has to be plugged in to the recorder at one end and the mains at the other. The end that connects to the tape recorder is strictly speaking a socket (because it would be dangerous to have the live pins of a plug sticking out). The recorder in this case is fitted with a recessed plug to take the socket in the end of the lead. This recessed plug is usually fitted near, but not actually on, the input/output socket panel.

Alternative high impedance output

Grundig models on which external connexions are made by three-pin plug offer an alternative high impedance output from the diode socket. The diode input is connected to pins 1 and 2 (looking on the face of the socket) and the high impedance output to pins 3 and 2. Used in conjunction with connecting lead S.L.233 and J3 plug, the socket connexions allow a recording to be made from a Grundig radio set diode socket and played back (via the same lead) through the radio amplifier and speaker by simply switching the recorder to playback, when the output lead is connected to the radio pickup input socket.

CHAPTER 2

The Grundig Models

SINCE THE FIRST GRUNDIG TAPE RECORDER
was marketed, there have been well over 30 different models
and many of the original machines are still giving good
service. There is no room to describe these early models in
detail in this chapter, but the Data Section gives all the
principal particulars and brief working instructions for
every Grundig model since the first.

Broadly speaking, all the models can be grouped first as
two or four track machines and then divided into monaural
and stereo. The battery operated portable machines form a
special group on their own. Which model you choose depends a lot on the sort of things you want to do with it, so it
is worth while to know what the differences are.

Two-Track Models

A two-track tape recorder makes two separate half-track
records on one tape. It has a single record/play head that
records a track slightly less than half the width of the tape
(about 1/10 in.) and it records it along the upper half as the
tape moves through the sound channel from the supply
spool on the left to the take-up spool on the right.

When all the tape has been recorded in this way and it has
all been wound on to the right hand spool, you change the
spools over, putting the full spool on the left and the empty
one on the right. But as you transfer the full spool, you also
turn it over. When you do this, you bring the blank half track
to the top and the half-track you have recorded goes to the
bottom.

You now repeat the recording process, and this time the
record goes on to the blank half-track. In this way you put

32

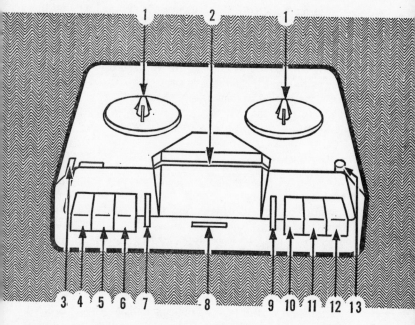

TYPICAL "POPULAR" 2-TRACK MODEL (TK14 etc.) 1 spool holders. 2 sound channel. 3 Tape position indicator. 4 Fast Wind, right to left. 5 Pause. 6 Mic/Radio, P.U. input selector. 7 On/off/Tone. 8 Magic Eye. 9 Recording level/Volume. 10 Start. 11 Stop. 12 Fast Wind. left to right. 13 Record.

two records on the tape, each occupying rather less than half the width.

When you want to play back the tape, you can select either record at will. There is only one point to remember. If the half-track you want to play is on top when the full spool is in the playing position on the left, you can go straight ahead and play it. But if the half you want to play is underneath, you have to wind all the tape on to the other spool (using fast forward wind, p. 16) and then change over the spools. This will bring the half-track you want up to the top.

Alternatively, you can transpose the spools first—putting the full spool on the right, and wind the tape back on to the empty spool (using fast reverse wind). The tape will then be ready to play without any further switching around.

You can make things easier for yourself if you remember always to wind the tape back on to its original spool after you have played it. This way you know that every spool is ready to play the first record without rewinding, and you have to re-wind it to play the second record.

Four-Track Models

A 4-track model makes four separate quarter track records on a single tape. It has two record/play heads mounted one on top of the other, each recording a track less than one quarter the width of the tape (about 1/25 in.). The upper head records along the top quarter of the tape (Track 1). When this head has recorded the whole length of the tape and you switch the spools over and start again, it will record along what was the bottom quarter of the tape (Track 2). So you now have two quarter-track records on the outside of the tape—Tracks 1 and 2—and two blank quarter-tracks on the inside—Tracks 3 and 4.

At this point you switch off the upper head and switch on the lower one. There is a quarter-track space separating this head from the top head, so this one records the quarter-track immediately below the centre of the tape (Track 3). You now record this track along the full length of the tape and then switch the spools around. This time you bring the remaining blank quarter track opposite the lower head. When you have recorded that (Track 4), you have four records on the tape; two of them (1 and 2) can be played with one head and two (3 and 4) with the other. You select the track you want to play by first turning the track selector (p. 35) to the appropriate position marked 1–2 or 3–4.

Next you have to decide whether you can play the record you want straight off the spool, or whether you will have to rewind the tape first (as for Track 2 of a half-track recording, above). This is easy so long as you always return the tape to

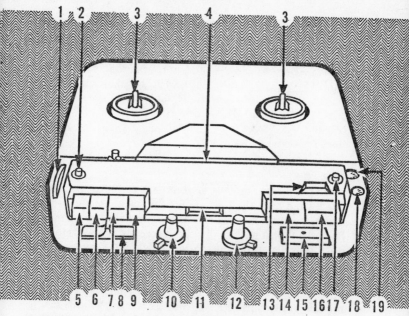

TYPICAL "LUXURY" 4-TRACK MODEL (TK40 etc.) 1 Speed Selector. 2 Tape cleaner release. 3 Spool holders. 4 Sound channel. 5 Record. 6 Trick (Superimpose). 7 Stop. 8 Fast Wind. 9 Start. 10 Monitor volume/Speaker On-off. 11 Magic Eye. 12 Input Selector/Recording Level/Volume. 13 Tape position indicator. 14 Track 1-2 selector. 15 Pause. 16 Track 3–4 selector. 17 Tape indicator reset. 18 Cine sound head input. 19 Microphone input.

its original spool and start from there. You will be able to play Track 1 and Track 3 by simply setting the track selector to the right number, but if you want to play Track 2 or Track 4, you will have to rewind the tape first.

For and Against

If you are now wondering what the differences are between these two groups of models, here they are:

Half-track recording, since it makes use of a bigger area of the magnetic coating, gives you better quality, but it uses more tape. Quarter-track gives you twice as much recording time for your money, but, since it can use only half the amount of magnetic material, cannot give you the highest quality. However, you need to have a very good ear for quality—or be using a first rate external sound set-up (p. 131)—to tell the difference, so go for tape economy and choose the quarter-track model unless you are a real high fidelity enthusiast. However, that isn't the whole story, because the 4-track machine still has a trick or two up its sleeve:

Two Tracks at once

On the 4-track models, if you make a record with the top head, wind back the tape, switch the track selector over and make a second record with the lower head, you can wind back the tape and then play both records at once. How you do this will depend on the model. Some 4-track Grundig models have a slider on the deck with three positions: Tracks 1–2, Tracks 3–4 and D. When you set the slider to D, both tracks—1 and 3 or 2 and 4—play back together. Other models use two press keys to select the track—one key for Tracks 1–2 and the other for Tracks 3–4. If you press both these keys down together, you hear both records at once.

Being able to play two tracks together in this way is useful when you want to add a musical background or sound effects to a speech record and so on. By noting the reading on the tape position indicator at the start and finish of a particular item, you can add the music on the second track so that it fits in reasonably well, but you cannot synchronise the second record exactly with the first, because you can't actually listen to the first record while you are making the second. However, there is a way of doing just this:

Two Tracks in Synchronism

On the 4-track models, when you set the track selector to record from one head, the other head is automatically connected direct to the high impedance output socket. You can connect a small transistorised preamplifier (p. 156) to this output and boost the signal so that you can hear it with a

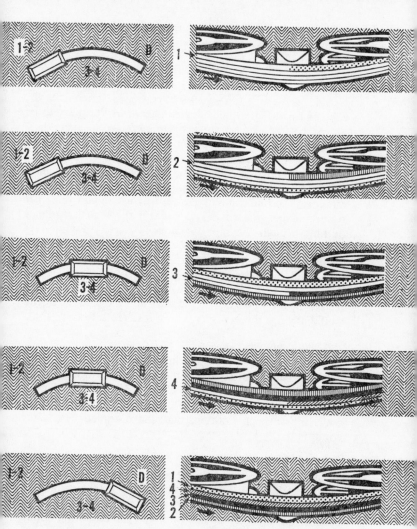

FOUR-TRACK RECORDING. 1 To record Track 1, set track selector to 1-2. 2 To record Track 2, turn spools over and interchange. 3 To record Track 3, trun spools over and interchange. Set track selector to 3-4. 4 To record Track 4, turn spools over and interchange. 5 To play Tracks 1 and 3 or 2 and 4 together, set track selector to D.

stethoscope earphone plugged into the output socket of the preamplifier and make your next record in synchronism on the other track.

Playing the Second Stereo Channel

A 4-track stereo tape (p. 87) carries two stereo records, each consisting of a pair of recorded tracks that have to be played back through two separate amplifier/speaker units. If you play a 4-track stereo tape on an ordinary monaural 4-track model, the built-in amplifier and speaker will reproduce one of the stereo channels, just like a normal monaural recording, but it has no provision for amplifying and reproducing the second channel. However, you can play the other stereo channel record if you connect the high impedance output through the monitor preamplifier as if you were going to synchronise two recordings (above), but this time instead of connecting an earphone to the output of the preamplifier, you connect an external power amplifier and speaker, e.g. the Grundig CR1 (p. 152). You can then get the stereo effect by placing the recorder and the speaker as you would arrange the two speakers of a stereo set-up (p. 47).

Automatic Models

On these models the recording level adjusts itself automatically and there is no recording level indicator. The TK23L and TK23A can be switched to automatic or manual control at will. There is a recording level indicator on these models which operates in both manual and automatic recording conditions.

Stereo Models

For stereo sound, you need two complete recording and playback channels. You start by picking up the sound you want to record on two separate microphones, then you feed each of the separate signals into separate recording amplifiers and record them with two heads on two tracks on the same tape. In the same way, when you want to play the tape back, you use two separate playback heads, feeding into two separate amplifiers to drive two separate speakers. By

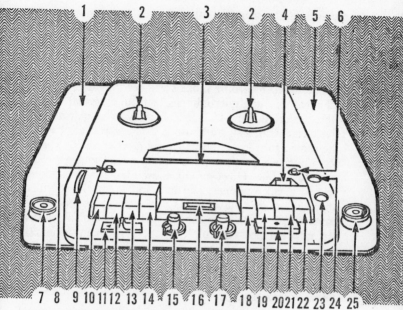

*TYPICAL STEREO MODEL (TK46). 1 Left hand channel speaker.
2 Spool holders. 3 Sound channel. 4 Tape position indicator. 5 Right hand
channel speaker. 6 Tape indicator reset. 7 Cross record/Record/Erase
head cutout. 8 Tape cleaner release button. 9 Speed selector/On-off.
10 Record, Tracks 1&2. 11 Fast Wind. 12 Record, Tracks 3&4. 13 Stop.
14 Start. 15 Bass–Treble tone controls. 16 Magic Eye. 17 Input selector,
18 Tape monitor. 19 Synchronous monitor. 20 Pause. 21 Playback,
Tracks 1 & 2. 22 Playback, Tracks 3 & 4. 23 Microphone input, R.H.
Stereo channel. 24 Microphone input, L.H. Stereo channel and Mono.
25 Twin volume controls for RH and LH speakers.*

listening to the speakers at a point equidistant from each,
your ears hear the sound as if it came to them from the
original source—sounds recorded from the left hand side
appear to come from that side, sounds recorded from the
centre appear to come from between the speakers, and so on.
The effect is much more vivid than the best single channel
(usually called monaural) reproduction can achieve.

There are two types of Grundig stereo model. One will
play back pre-recorded stereo tapes, but it cannot make its

39

own stereo records. The other type will both record and play stereo sound. Both types have a stereo/mono switch which lets you change over from stereo to monaural working at will. The 2-track models can record only one stereo record on the tape, the 4-track model can record two.

The stereo recording models will record either from two identical Grundig microphones or from a special Grundig stereo microphone which incorporates two separate sensitive elements in a single housing. The controls are basically the same as those on the monaural models, the two sets of recording level, volume, tone and so on being ganged to operate as one. However, to allow for variation in room acoustics, microphones, and in amplifier/speaker performance, all models have some provision for adjusting the balance between the two channels. On some models, this is a separate balance control, on others you can adjust the volume from each channel independently and then lock the two controls together.

Speaker arrangements

The speaker arrangements vary according to the model. On one model, the two speakers are clipped, one on each side of the main stereo recorder. They are intended to be unshipped and spaced out at the end of long extension leads. Later models have the speakers permanently built in on each side of the deck. The stereo effect is reduced by mounting the speakers as close as this, but the arrangement is self-contained and portable. In addition, there are low and high impedance output sockets for each channel, so that you can mute the speakers on the recorder and play the tape either through two extension speakers connected to the low impedance outputs or through high fidelity stereo sound equipment connected to the high impedance outputs.

Special facilities

The current stereo model records two quarter-track stereo records on the tape. It has separate recording and playback heads, each one with its own amplifier instead of the usual

40

combined record/playback arrangement. This means that you can play back one track while you are recording the other and you can also cross record—feeding the playback output of one channel into the recording input of the other.

You can also listen with the playback head to the recording as you are making it on the tape, the playback head picking up the signal off the tape a fraction of a second after the recording head has recorded it. This gives you a running check on the actual quality of the record you have made instead of the signal you are recording.

Finally, with this model you can make use of the separate record and playback heads to create an echo effect. To do this you set the controls to feed back a part of the output from the playback head into the input of the recording head. Because the tape reaches the playback head a small fraction of a second after it passes over the recording head, the signal you feed back lags slightly behind the original signal. It goes on feeding back again and again, getting weaker each time until there is nothing left, exactly like an echo. You vary the time and quality of the echo according to the tape speed you select—a slow tape speed creates a long time lag between the original sound and the echo; a high tape speed gives a quicker, livelier effect.

Battery models

These models are self contained and are designed primarily to operate from their own built in dry battery power so that they can be used out of doors and away from the normal mains electricity supply. However, all models can also be operated from the mains when necessary to economise in battery power. On the earlier models you had to replace the batteries with a detachable mains power pack. One of the recent models incorporates the mains power pack as well as the battery. On this model the mains lead and its plug are stowed in a recess in the side of the recorder; when you remove the plug to connect it to the mains you automatically disconnect the batteries and switch over to the mains power pack.

41

The size and weight of these models has been greatly reduced by the use of transistorised printed circuits, but the early models although small and light, had only one tape speed, there was no fast forward wind, no tape position indicator and only 3-inch spools. The current models are high quality machines with most of the facilities associated with the mains operated types.

All the current portables have a high impedence output socket for playing the tape back through external amplifying equipment and some have an extension speaker socket.

Battery life on these machines varies widely according to the kind of use they get—intermittent working gives a longer total life and continuous running shortens it. With average use—say half an hour a day, the batteries should last for about 15 to 20 hours before they need renewing.

Cassette Models

The first tape cassette model used the DC International cassette, giving 2 hours total playing time at a speed of 2 i.p.s. The latest model uses Compact cassettes, giving total playing times of 60, 90 and 120 minutes at $1\frac{7}{8}$ i.p.s. This model will also play mono or stereo pre-recorded Musicassettes.

Automatic models

On these models the recording level adjusts itself automatically and there is no recording level indicator. Automatic models are much simpler to use and adjust the recording level much more accurately than it is possible to do it by hand with the majority of subjects. Some have a recording level indicator and optional manual control.

De Luxe models

In 1965 Grundig introduced a series of new de luxe models in distinctive styling and finishes. These models are distinguished from the models on which they are based by the letter L—e.g. TK23L. Subsequent models following the same styling drop the L.

CHAPTER 3

How to Use your Grundig

BEFORE YOU CAN RECORD OR PLAY BACK
with your Grundig you must connect it to the main electricity supply and load it with the tape.

To connect your Grundig to the mains

You connect your Grundig to the electricity supply
through the lead supplied with it. In models which have
three-core cable, you should plug into a 5 amp power (three
pin) socket. When not in use, the lead is stored in a pocket
next to the distributor panel or under a flap on the deck, or
sometimes in a compartment provided in the lid of the
recorder.

Remember that your Grundig is designed to work off an
A.C. supply. Unless you have a special converter (see p. 62)

*LOADING THE TAPE. 1 Place full spool on left spindle, tape leading
off left of spool. 2 Drop tape into tape slot in sound channel.*

you must never connect it to a D.C. supply or you will seriously damage it. Before you connect it to the A.C. supply you must see that the mains adjuster is set for the correct voltage and that you have the correct plug.

Grundig tape recorders are made to work off either 50 or 60 cycle supplies. The two are not interchangeable; if you connect your machine to a supply of the wrong frequency it will not run at the proper speed and you will not get satisfactory results.

To load the tape

Loading the tape on your Grundig is easy. Before you start recording or playing back you fit a full spool of tape on the left and an empty one on the right, like this:

1. Place the full spool on the left-hand holder with the end of the tape coming towards you off the left of the spool.

2. Turn the spool until you feel the projections on the spindle fit into place.

3. Fit an empty spool on the right-hand side in the same way.

4. Pull off about 12 in. of the leader tape and lower it into the tape slot in the sound channel cover. Make sure the shiny side of the tape faces you as it comes off the spool and that there are no twists or kinks.

3 Draw end of tape through slit in flange of empty spool. 4 Turn spool until tape overlaps itself and then continue until auto-stop foil is clear of channel.

5. Pull the tape through the sound channel until you can draw the end into the slot in the flange of the empty spool, leaving an inch or so of tape to grip at the hub.
6. Hold on to the end of the tape and turn the spool in an anti-clockwise direction until the tape winds on to itself.
7. Go on turning until you have wound the metal foil insert on the tape on to the right-hand spool (otherwise the automatic stop will operate as soon as the foil reaches the sound channel).
8. Set the tape position indicator to zero.

The tape is now ready to record or play back.

On the 2-track models, if you want to use track 2 first, you have to fast wind all the tape on to the right-hand spool and then change over the spools. On the 4-track models the same thing goes for tracks 2 and 4.

Note. On the cassette loaded battery portable model you simply fit the cassette in position; it is not necessary to thread the tape between the spools.

How to Record

Whether you want to make a record from your microphone, radio, gramophone pick-up or any other signal source, the principle is the same.

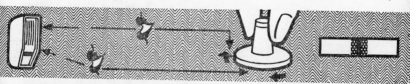

RECORDING YOUR VOICE. Speak fairly close (1 ft.) to microphone with the signal level control normal. If you speak at a distance and have to set the level control high to get a good reading on the magic eye (right), you will exaggerate background noises and lose quality.

1. Connect input and set input selector.
2. Select speed.
3. Select track.
4. Switch on.
5. Switch off monitor speaker.
6. Set tape position indicator to zero.
7. Press record key.
8. Adjust recording level.
9. Press start key.

45

10. Make your record.

11. Press stop key.

Notes on the above:

2. If you are recording your voice, the slowest speed will do; for high quality music, use the highest speed.

5. If you are recording through the microphone, and forget to switch off the speaker, the sound from the speaker will feed back into the microphone and build up into a loud howl. For recording from radio, pick-up, etc., you can use the monitor speaker to check your recording as you go along.

7. On some earlier models the tape starts when you switch to record. With these machines you should press the pause key to halt the tape until you have set the recording level.

8. Most models have a magic eye or meter indicator, but the automatic models have no indicator; with these the recording level is set automatically and you can omit this step.

   To adjust the level on all other models you turn the recording level knobs up or down until the two segments of the magic eye just come together on the loudest sounds.

How to Wind Back

When you finish making the record, all the recorded tape will be on the right hand spool and you have to wind it back

HOW TO USE THE MAGIC EYE. Left: *Segments do not meet on loudest signals. Recording level too low. Advance control.* Centre. *Segments almost touching on normal signal. Recording level too high. Turn back control.* Right: *Segments just meet on strongest signals. Recording level correctly adjusted.*

again on to the supply spool on the left before you can hear what it sounds like. This is how you do it:

1. Engage the fast reverse wind control.

2. When all the tape is back on the left hand spool, press the stop key.

Notes on the above:

1. Before you start the tape running back at speed, turn one of the spools by hand to take up any slack in the tape. If there is a lot of

46

slack it may break with the sudden jerk of starting or get snarled up in the sound channel.

2. On models which have an automatic stop, the spools will come to rest as soon as the stop foil on the tape runs through the sound channel. This will automatically return the control to neutral.

How to Play Back

When you have wound back to the beginning of your record and stopped the tape, all press keys will be up and the motor will be running, but the tape will be at rest. You can now play back your record—through the built-in speaker of your Grundig—like this:

1. Press start key.
2. Adjust volume.
3. Adjust tone.
4. Press stop key at end of record.

Notes on the above

1. On some early models, you have to set a control to Play first. On current models pressing the start key automatically plays the tape unless you have first pressed the record key.
3. The best setting of the tone control is usually the one that gives you as much 'top' as possible and stops just short of the point where you can hear a faint hiss on the quieter parts of the record.

Other ways of playing back

In addition to playing back normally through the built-in speaker, there are several other ways of listening to the record on the tape.

Playing back stereo tapes

Stereo tapes carry two records side by side which have to be played back through two separate speakers. One stereo model reproduces one channel through a built-in power amplifier and speaker, and you connect an external amplifier and speaker—*e.g.* the Grundig CR1 unit—to the high impedance output from the other channel. On another model each channel has its own amplifier/speaker unit and a removable speaker. In each case you place the speakers about 6 ft. apart and facing you. The latest stereo model has two built-in speakers and facilities for playing through either two extension speakers or two channel high fidelity equipment. The combined effect is the nearest possible approach to the three-dimensional quality of the original sound. You can play back either commercial pre-recorded tapes or tapes recorded on a Grundig stereo model (p. 38).

Playing through an external amplifier

If you are a real enthusiast you may already have an independent amplifier connected to its own high fidelity speaker. You can play back your Grundig tapes through this type of equipment and get the full benefit of the high standard of the recording. (You can never achieve this from even the best of internal speakers since something has to be sacrificed in the interest of portability.)

All you need is a spare screened lead connected from the high impedance output socket at the back of your Grundig to the input of the amplifier. When you do this you may silence the internal speaker or speakers on your Grundig with the appropriate control indicated on the individual model data sheet at the back of the book.

If you have a really good console radio or radio gramophone you can make use of its amplifier and speaker in the same way. Here again you connect the high impedance output socket on your Grundig through a screened lead to the pick-up socket on the back of the radio set. If you want to play back through a radiogram it is unlikely that there will be a socket for an external pick-up, but it is a simple matter to connect up a suitable socket to the amplifier power stage; if you cannot manage this yourself, your Grundig dealer will do it for you.

Playing through an extension speaker

The bass reproduction of your Grundig is limited by the size of its built-in loudspeaker. By connecting it to a large extension speaker you can enjoy more of the rich lower frequencies recorded on the tape. As practically all extension speakers are of low impedance (3, 7 or 15 ohm), all you have to do is to connect the low impedance socket to the larger speaker with a suitable length of screened lead. You will need to mute the internal speaker; with most models the tone control remains effective. If you like you can have your Grundig in one room and the extension speaker in another.

Grundig Model TK41 has an output of 7 watts when connected to an extension speaker, but the power is cut down when the internal speaker is in use.

Using the Grundig as an amplifier

All portable tape recorders incorporate an amplifier for playing back the signal off the tape. This amplifier is similar to the independent units used for amplifying the signals from microphones for public address and from a gramophone pick-up or radio tuner. Some Grundig models incorporate switching to use their amplifiers this way.

On some earlier models, a single key is provided for this purpose; on some two-way decks you have only to depress both track 1 and track 2 keys and release them together and you have a 'straight through' circuit from the selected input, through the amplifier to the outputs.

This straight through facility is useful when the recorder is used in conjunction with a radio tuner or gramophone pick-up, and perhaps an extension speaker. For ordinary listening you switch the amplifier straight through from the radio tuner or pick-up. If an item comes along that you want to record, you simply press the recording key—remembering to check the input level if only a single dual purpose control is fitted (your playback volume setting may not be suitable also for recording in this case). On many Grundig models, you can still continue to hear the programme as you record it—either through headphones or the loudspeaker. This is called monitoring. (p. 28)

By using the 'straight through' amplifier facility in this way you can enjoy first class radio listening for the extra cost of a radio tuner instead of a complete radio set. And for disc reproduction you need only buy the pick-up and turntable unit instead of the whole record player.

Doing things properly

While it is fun to take the Grundig around with you and make recordings here, there, and everywhere a great deal of its life will be spent at home making records from the radio and playing back tapes for your entertainment. So it's worth while to arrange a permanent place for it in the room where you normally use it. This place should be close to your radio and gramophone and the supply point, so that you can keep your electrical connexions as short as possible. (The shorter you can keep the trailing cables, the less likely they will be to get snarled up and the less risk there will be of picking up stray hum.) Always use the correct Grundig leads and plugs and don't take chances with makeshift connexions, because an accident or short circuit can be an expensive business.

You can make things very much easier for yourself during long recording or playback sessions if you can have your Grundig on a low stool at the side of your easy-chair.

Recording from other sources

Your microphone is only one of the many signal-makers that you can record on tape. There is just as much fun and interest to be had from recording radio programmes, 'dubbing' from a second tape recorder, and so on, as the following pages will show you.

Your radio set

You can make records of programmes from your radio as easily as from your microphone; in fact many Grundig owners regard this facility as being more important than any other. Once you get the hang of it you can build up a library of recorded items that will be a source of endless enjoyment.

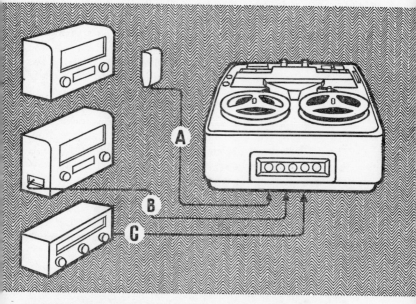

RECORDING FROM RADIO. You can record with the microphone in front of the radio loudspeaker (A) or by connecting the radio extension speaker to the radio L.S. input (B) or by connecting the radio diode or a radio tuning unit to the diode input (C). C gives the best quality, B is almost as good, A is not recommended for high quality recording.

First of all you have to remember that your Grundig is going to record everything, including interference, so if your radio is not up to standard, your recordings will be no better. If you want your recordings to be free from interference, atmospherics and background noises then you will have to get an FM radio set or tuning unit. FM (frequency modulated) radio signals are practically free from the noises and instability that affect an ordinary AM (amplitude modulated) radio. (FM radio is also called VHF—short for Very High Frequency).

51

If you decide to switch over to an FM receiver, remember that it is not necessary to buy a complete set. Every radio set consists of two sections—the radio frequency section and the amplifier. The radio end of the set, which you can buy separately as a Tuning Unit, receives the weak high frequency impulses from the aerial and converts them into an audio frequency signal. This signal is passed on to the amplifier where it is amplified sufficiently to work the loudspeaker. However, on all Grundig models you already have a suitable amplifier in your tape recorder through which you can play a tuning unit—and so long as you don't mind using the recorder every time you want to listen to the radio, you need only buy the radio tuning section. There is a very wide selection of suitable radio tuning units that you can connect up to your tape recorder in this way. The arrangement has two advantages: (1) you save the cost of an extra amplifier and loudspeaker, and (2) whenever you want to record anything coming over the radio, you are all set up to do it at the turn of a switch.

There are four ways of recording a radio programme on your Grundig: (1) through the microphone, (2) from the radio extension speaker output, (3) from the radio set diode, and (4) from a separate radio tuning unit. With methods (1) and (2), if your radio has a special bass lift control, turn it right down before you start because your Grundig can provide all the bass compensation necessary.

RECORDING RADIO THROUGH THE MICROPHONE. The easiest way of taking down items from radio is to put your microphone between 1 and 2 ft. in front of the speaker of your radio set, adjust the radio tuning and tone control to give the best possible quality and about normal listening volume, and record exactly as for a normal microphone subject.

While this is the easiest way, it is also the least satisfactory. For one thing, the microphone will pick up all the room noises—footsteps, doors closing conversation, etc.—and all the echoes and reverberations. In addition, the recording will include all the imperfections of both loudspeaker and microphone and any shortcomings of the amplifier in the radio set (which always exist to some extent). So you should only use this method in an emergency or when the quality of the signal is not important—*e.g.*, for speeches and talks.

RECORDING FROM THE RADIO EXTENSION SPEAKER OUTPUT. For this method you by-pass the loudspeaker and microphone in the previous set-up and connect the extension speaker output of the radio set directly to the input of the tape recorder. The radio speaker remains operative and enables you to monitor the signal. You will need a connecting lead fitted with plugs at each end to suit the radio output and your Grundig input. Now proceed as follows:

1. Connect radio extension loudspeaker output to the radio L.S. input with a screened lead.
2. Tune radio to programme required.
3. Set radio volume to normal listening level (and turn down bass lift control if fitted).
4. Now proceed as for recording from microphone, but select radio L.S. input instead of microphone.

With this method the results will be vastly better than you get from a direct microphone recording, and if the amplifier of your radio set is of reasonably good quality you may be completely satisfied with the result. As a rule, however, this end of the radio set is only just good enough for its job, and is not really up to the standard you want if you are going to play back your tapes through high fidelity equipment.

RECORDING FROM THE RADIO DIODE. You get the best quality if you connect your recording lead to the radio circuit at a point just before the amplifier. This means that you connect to the diode which is the valve or other type of rectifier that turns the energy received from the aerial into a usable signal. If you know enough about radio circuits to follow the diagram on p. 54 and make the necessary connexion, well and good, but if you have any doubts about your ability, you had better be on the safe side and ask your Grundig dealer to do the job. In any case you will want to bring the connexion out to a socket fitted at a convenient

53

point in the back of your radio set; or you can connect directly to a suitable length of screened lead with a plug on the end for the input of your tape recorder.

The method of recording is the same as above:

1. Connect radio diode output to the diode input with a screened lead.
2. Tune radio to programme required.
3. Set normal radio volume if required for listening or turn down fully.
4. Now proceed as for recording from microphone, but select diode input instead of microphone.

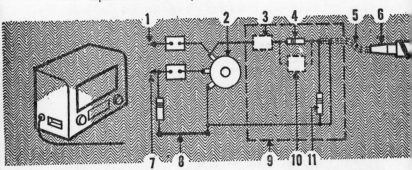

RECORDING FROM RADIO DIODE CONNEXION. By recording direct from the diode of your radio you can get results just as good as from a radio tuner because you by-pass the radio amplifier (usually of only moderate quality). This is how to make the diode connexion. (For technical types only; others should get their Grundig dealer to do it.) 1 From diode. 2 Volume control. 3 ·025 μF condenser. (Insert if D.C. is present on volume control.) 4 1-2 megohm resistance (not critical). 5 Screened lead from radio set to tape recorder. 6 Plug. (Insert in diode socket of Grundig.) 7 To audio stages of radio. 8 Chassis. 9 Broken line encloses additional items. 10 35 PF condenser. (Add if long connecting lead is used.) 11 50,000-100,000 ohm resistance (not critical).

RECORDING FROM A RADIO TUNING UNIT. A radio tuning unit is, in effect, a radio set with the amplifier and speaker taken away. It is no use without a separate amplifier unless you only want to use it for recording on your Grundig.

You can get both AM and FM tuning units. An AM unit will bring in a wide range of stations, both near and far, but it is subject to interference; an FM unit restricts you to a few nearer stations but it gives you a clean signal completely free from noise and interference.

In principle there is no difference between making a record from the diode of an ordinary radio set and from the output of a radio tuning unit. So the recording drill is pretty much the same for both:

1. Connect tuner unit output to the diode input with a screened lead.
2. Tune to programme required.
3. Set volume control of tuner unit for normal listening through your Grundig speaker.
4. Now proceed as for recording from microphone, but select diode input instead of microphone.

If you want to go on listening without recording, you can set the controls on some Grundig models to convert them into a straight-through amplifier, so that you can listen to the programme either through the built-in speaker or through external sound equipment.

And of course you can record in silence if you turn down the Grundig volume control and the one on the external sound equipment.

This method of recording radio programmes is by far the best and if you have a good FM tuning unit and aerial you will be able to make recordings at least equal in quality to commercial pre-recorded tapes or discs.

Your gramophone

There are three ways of using your Grundig to transfer a record from discs on to tape: (1) through the microphone (using a radiogram or record player to reproduce the disc), (2) from the extension specker socket of a radiogram or record player, and (3) direct from the pick-up of the record turn-table. These three methods correspond to the methods for recording from a radio.

Here again the easiest method but the one that gives the poorest quality is to record through the microphone. Next in order of quality come recordings made by connecting to the extension speaker socket or terminals. But if you want

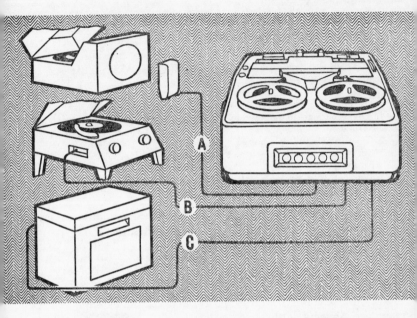

the best possible quality you must choose the third method and record either direct from the pick-up (via a suitable equalising circuit) or from the pick-up via a pre-amplifier (p. 132).

The Radio L.S./Gram.P.U. socket on your Grundig is suitable for connecting either direct to a gramophone crystal pick-up or to a low impedance extension speaker output on an A.C. mains model radiogram or record player. If your disc playing equipment has no provision for feeding an extension speaker or is designed for a high impedance speaker, or if it is of the A.C./D.C. type, you should consult your Grundig dealer and get him to fit a suitable output connexion.

Recording from Radiogram or Record Player. If your radiogram or record player has a control for boosting the bass frequencies, turn it down as far as it will go before you start and only use it if tests show it is an improvement.

Here is the way to make your record:

1. Connect the pick-up or extension speaker output socket with a screened lead to the correct input on the recorder.
2. Switch the controls to record so that the recording level can be adjusted.
3. Start the gramophone turntable and lower the pick-up on to the record.
4. Adjust the recording level as needed.
5. Remove the pick-up and start the tape.
6. Re-start the pick-up at the beginning of the record.
7. At the end of the record, turn down the recording level and then stop the tape.

When recording from a radiogram or record player, the speaker will remain in circuit and you can use it for monitoring the signal. In this case it does not matter whether you have the Grundig speaker switched on or not. It is generally better to turn down or cut out the Grundig speaker with the output volume control. On some models the speaker is cut out automatically when you record.

Recording from Pick-up. If you are recording from a pick-up direct or through a pre-amplifier it is far better always to monitor the signal through the speaker of your Grundig. You can do this on most models, but not on the models where the speaker is automatically cut out in the record position. With these models, however, you can connect the pick-up socket of any radio set to the high impedance output socket and listen to the disc through the radio speaker.

If you have a radiogram, your Grundig dealer can modify the circuit so that you can record direct from the pick-up. You can also connect the output socket of your tape recorder to the power amplifier of the radiogram so that you can play the disc you have recorded through the speaker of your radiogram.

You should always play back the tape recording right away so that there is still time to have another try.

With the stereo models, of course, you have two inputs to connect from the pick-up.

Recording from a second tape recorder

A second tape recorder can be used for 'dubbing' all or parts of a record from one tape to another and for editing and creating special effects. You don't need to own the extra machine; you are almost certain to know another tape recorder enthusiast who will co-operate on a mutual exchange basis. Save up your dubbing jobs and settle down to clear them off in a single session when you can borrow the second machine. You may lose a little quality in transferring a record from one tape to another, but with two high quality machines the difference should not be noticeable. So make sure that the machine you borrow is up to the quality of your Grundig.

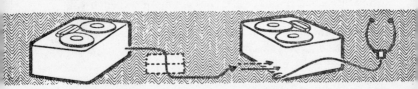

DUBBING FROM A SECOND RECORDER. The tape on the recorder (left) is played back normally while the output is connected to the input of the recorder (right) either direct (Output 1 to Radio L.S./ Gram. P.U. or Output 2 to Diode/T.A.), or through a mixer or pre-amplifier (output to Diode/T.A.). While the record is being transferred to the recorder right, signal may be monitored through headphones connected to Output 2.

First of all you will have to make up a screened lead with three-pin or jack plugs to suit your Grundig Radio L.S./ Gram. P.U. input at one end and the high impedance output socket of the second recorder at the other (p. 29).

Once you have the two machines connected with the screened lead the details of the recording procedure will depend on which machine is playing back (machine A) and which one is recording (machine B), but the general principle is as follows:

1. Switch on both recorders.
2. Load the tape to be dubbed on to machine A and a blank tape on to B.
3. Set A to playback and B to record.
4. Set the A output volume control to between 1/3 and 1/2 maximum.
5. Start A and adjust the recording signal level on B. (If you can't get it up to a satisfactory recording level, turn up the A output volume.)
6. Wind the A tape back to the start of the recorded section you want to transfer and turn down the volume control.
7. Start B, pause for a second or two and then start A, turning the volume control up to the previous level at once.
8. At the end of the recording, turn down the B recording level control and stop both machines.

While you are making the record you may be able to listen to the signal from the speakers of one or other machine at will, depending on the control facilities, or you may prefer to monitor the recording by watching the recording level indicator on machine B.

The job is made very much easier if you can do the dubbing through a Grundig mixer unit (p. 151). This will enable you to monitor the input signal with a pair of Grundig stethoscope earphones and it will also allow you to record a commentary or add music or effects to the recording as you go along.

The telephone

You can make a record on your Grundig of both sides of a conversation on your office or private telephone. For this you use the Grundig Telephone Adaptor (p. 154) which plugs into the diode input socket.

The business end of the telephone adaptor lead is a coil in a plastic case. The case is fitted with a rubber sucker so that you can stick it on to the side of your telephone base. The coil in the adaptor responds to the magnetic field set up around the coils in the desk unit when you are having a telephone conversation. The coil will pick up both sides of the conversation equally well, but its position on the base is important.

The first job is to find the spot on the base where the coil picks up the strongest signal. Here is how you do it:

1. Select the correct input and plug in the adaptor.

2. Set the Grundig controls to the record position but do not start the tape.

3. Turn down the speaker volume control.

4. If you are on an automatic exchange, simply lift the telephone receiver off the hook and use the dialling tone for a test signal. If you are on a manual exchange, ring up one of your friends and ask them to count or recite the alphabet.

5. Hold the telephone adaptor just clear of the telephone base and move it about while you listen to the test signal.

6. Once you have found the spot where the signal is strongest, moisten the sucker and stick it in place.

On Grundig models where the speaker is automatically cut out by the record switch you will have to find the right spot by watching the magic eye. In this case you want to find the position where the segments of the indicator just

RECORDING FROM TELEPHONE ADAPTOR. Both ends of telephone conversation set up magnetic vibrations around transformer coils in base of instrument. Telephone adaptor coil converts these vibrations into an electrical signal which is amplified and fed (a) to recording head, (b) to Grundig speaker so that several people can hear the conversation being recorded.

close on the strongest signals with the signal level control set about ½ way.

Once you have found the best position for the coil you can go ahead and start the tape by pressing the start key, or releasing the temporary stop button, according to model. The recording procedure is the same as for recording from microphone.

You can notice a certain amount of hum when recording from a telephone attachment. This is because the coil is highly sensitive to stray magnetic fields. You may have the adaptor too close to the Grundig itself, thus picking up hum from the motor; if possible you should always place it 4½ ft. or more away. If you still notice hum after moving the tape recorder away the trouble is probably coming from the mains supply. You can usually cure trouble from this quarter by moving the telephone about the room *with the receiver on the hook* until the hum or the deflection of the magic eye is at its lowest value. Make a note of this position and always put the telephone there when you want to record from it.

Stereo recording

With the Grundig stereo models you can record your own stereo tapes and play them back. The actual recording principles are the same; you simply make two records side by side on the tape exactly as though you were recording a single track. There is only one magic eye to watch and one recording level control adjusts both channels.

The chief difference between stereo and normal single-channel recording is that you use two microphones, each one plugged into a separate recording input. Start by placing the microphones about 6 ft. apart and facing you. Next adjust the recording level and make a record of your voice first as you walk from one side of the room to the other, then up to the microphones and away again. Now play back the tape (p. 47) and study the result. You will soon find out the right microphone spacing and the range of movement that give you the best stereo effect. You can go on to use the knowledge to make amazingly lifelike stereo records of all kinds of noises, sound effects and musical performances. With these models you can also transfer stereo disc records on to tape—but don't forget the Copyright Act.

Running off a battery

Unless you own one of the special battery-operated models (p. 41) your Grundig is designed primarily to run off the normal A.C. mains supply. This is all that most people ever want to do. But if you go in for any special hobbies that take you off the beaten track and away from power supplies, you can still take your Grundig with you and use it both for making records and playing them back. To do this you need two things: a storage battery and a D.C./A.C. converter. The obvious thing, if you have a car, is to use the car battery for the job.

There are two types of converter: rotary and electronic. Both will run off your car battery and provide alternating current for your Grundig. However, the speed of the rotary type is more difficult to control and it tends to generate hum, so the electronic converter is the one to use. It is generally smaller, lighter and uses less current. Its speed variation is no more than 1-2% against 10-15% for the rotary type, and it costs no more.

CHAPTER 4

The Technique of Recording

NOW THAT YOU KNOW HOW TO GO THROUGH the motions of recording and playing back with your Grundig, you are ready to start doing something about it. Most people start by recording their own voices and then go on to record the voices of their family and friends. Then from there they go on to recording from other sources— radio, discs, a second tape recorder and so on. But voices and direct microphone recordings usually come first, and of these, your own voice is the easiest subject to start on.

Your own voice

Your own voice is the easiest subject you could choose and it has the advantage that you can do it without an audience. Also, if you want to do interviews and provide spoken commentaries and continuity for your other tapes, you will need to be able to make clear and confident records of your own voice—so it is worth spending some time on it.

Make a start by sitting at a table with the microphone on its stand about 18 in. away from your face. Have your Grundig on a low table or chair at your side; don't put it on the table beside the microphone and don't stand the microphone on top of it or it may pick up hum or mechanical noise from the motor.

First of all decide what you are going to say. Don't wait until you start the tape, or you are sure to dry up and end by talking self-conscious nonsense. If you are no good at talking 'off the cuff', practice with your Grundig will soon put that right; but in the meantime it is better to read a passage out of a book.

Get everything ready for recording, but press the temporary stop control; now recite the alphabet in your ordinary speaking voice. As you speak, turn up the recording level control until the level indicator is just on the point of overloading. Note the position of the control and turn it back to zero.

You are now ready to make your record. Release the temporary stop, turn the recording level control back to the mark you have just noted, and start talking. Don't talk directly at the microphone—talk past it or over the top of it so that your breath doesn't impinge directly on it. If you shoot your words straight at the microphone your voice will sound 'breathy'.

On many Grundig models you can hear the recording, if you wish, through the speaker as it is being made. For these you may have the speaker either switched off altogether or turned right down to avoid feeding the sound from the

HOW TO USE YOUR GRUNDIG TO RECORD YOUR VOICE.
1 Assemble microphone and plug into microphone input. 2 Plug mains lead into supply point. 3 Load with tape. 4 Turn on-off control until it clicks.

speaker back into the microphone so that it builds up a howl (p. 45). But when you are recording your own voice you might find it helpful to turn up the output volume control until you can just hear your voice coming from the speaker as you talk. When you do this you must be careful to keep the volume to a minimum, otherwise you may start a howling oscillation when you raise your voice. On some Grundig models where the internal speaker is automatically cut out when you set the controls to record, the extension speaker socket still operates; so if you connect it to an external speaker or an amplifier/speaker set-up you will still be able to listen to your voice as you record.

The placing of the microphone and speaker have a lot to do with the level at which feedback howling starts. If it is troublesome try changing the relative position and angle of the microphone and tape recorder.

When you have said your piece, turn down the recording level and stop the tape. Wind back to the start, switch to playback, start the tape, and turn up the volume.

You will now hear the voice of a stranger. The words will be yours, but the voice—never! In fact, you will be listening for the first time in your life to the voice that everybody else

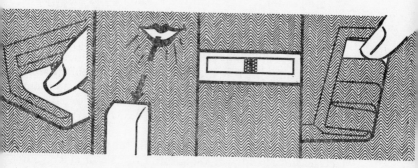

5 Set to record but do not start tape. 6 Start talking about 18 in. from microphone while you (7) adjust recording level control to correct setting by watching magic eye. 8 Start tape and commence recording.

hears when you talk. You never hear it yourself. The voice you hear reaches your ears mainly through the *inside* of your head—through bones and cavities that make it sound quite different from the voice that other people hear striking the *outside* of their ears. So don't blame your Grundig for faulty reproduction. You will soon realise how faithfully it reproduces when you hear your friends' voices. But if you don't like the sound of your own voice when you hear it for the first time, your Grundig will help you to improve it (p. 123).

Don't wipe out the record you have just made. Play it through once or twice and make a mental note of the things you would like to change. You may find that you were speaking too quickly or too slowly; perhaps you pitched your voice too high or too low; your diction may sound clipped or blurred.

Tackle the faults one at a time by repeating your test piece so that your second effort follows right after the first. Then play them back one after the other and compare them to see if you are getting any better.

It is worth while spending some time on this sort of exercise until you develop an easy, confident microphone manner and a clear pleasing tone of voice. Remember that you will generally be producer, sound engineer and one of the principal actors in every live recording you make. And if you are uncertain and nervous your voice will put a blight on the whole show.

All the time you have been recording and playing back your voice, you have been learning to handle your Grundig controls and to keep an eye on the recording level indicator. This is all part of the lesson, because it is most important that you should be able to do all these things automatically and without any technical hitches before you start bringing in other people.

There is nothing very difficult about all this. The manufacturer has already gone to a lot of trouble to make your Grundig simple to operate, and lots of people manage to work them quite well without ever having read a book about them (just as lots of people who have never read a book on photography can take snapshots). But a little extra trouble in learning the right way makes a big difference in the results you get.

When once you feel at home with the microphone and can operate the controls without thinking about them, you can make use of your skill in all kinds of different ways. The next step will be to record other people's voices.

Other people's voices

There are several ways of recording other people's voices: you may simply want them to record a message for your collection; you may want to record a conversation with them, interview fashion; you may want to make a candid record without letting them know.

In message and interview recordings, the other person knows what is going on and you can call on him to co-operate. In candid recordings you generally have to conceal both the microphone and the tape recorder and you can't expect any help from your victim.

HOW TO RECORD A MESSAGE. If the second person is accustomed to speaking into a microphone, your job is easy. Simply sit him in front of the microphone and get him to talk. Set the recording level control about halfway, and move the microphone closer or farther away until the magic eye shows the signal level is about right. You can now start making your record, keeping your eye on the signal level and holding its mean level steady with the control.

If the other person is nervous in front of the microphone you will have to spend a little time getting him over it. One way is to explain how the recorder works and encourage him to ask questions about it and make comments. Switch to record while he is doing this and then play the result back. This never fails to amuse the victim and start thawing him out. Then turn the microphone over to him and let him experiment. A few minutes of this sort of thing will get him into the right state for recording a straightforward message without drying up or sounding forced.

HOW TO RECORD AN INTERVIEW. This is one of the stock methods of getting information out of people. It is used constantly by reporters who work for radio and television programmes. So you have lots of opportunities to study how it should be done. And here is where you will see the value of putting in a little practice on your own first, because you have to be able to forget the recorder and the microphone and make the other person forget them too. If you have to fiddle with controls and microphone you will lose the intimate atmosphere essential to a successful interview.

For interviews you want the microphone to be unobtrusive. Don't place it directly between you and the person you are interviewing or you will both be conscious of it. Have it to one side or the other and as far away as you can place it and still get a reasonable signal. However, a lot will depend on the type of microphone you are using. Most Grundig microphones are omni-directional and will take in sounds all round; but some have directional properties and have to be placed more carefully (see p. 75).

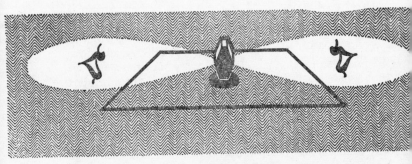

RECORDING OTHER PEOPLE (INTERVIEWS). A directional, ribbon type microphone (field shown white) is the right one for recording interviews. Noises from the sides are not recorded as strongly as voices from in front or behind.

The actual technique of interviewing is something that you will learn best by experimenting and taking notice of how the experts do it. You have to get the other person to do most of the talking with as little prompting from you as possible. Above all, don't ask point-blank questions that he can answer simply by saying yes or no. If necessary provoke him into an explanation by deliberately making a statement that you know to be wrong.

A good way of polishing up your interviewing technique is to take a few examples of broadcast interviews and study them. You won't need to use the interview technique very often unless you are a journalist or radio or newsreel reporter. Most of your records of family and friends will be of the message type. But there are sometimes personal items that you want to record which sound much better when you serve them up in this way.

How to make a Candid Record. Making candid records gives you all the thrill of stalking a quarry, and the results usually become show pieces that never lose their delight. You can practise this sport at any time—when you have a visitor, or when the children are engrossed in a game, or when the daily help stops for a gossip. And if you want to build up an unaffected sound picture of youngsters as they grow up, then candid records are the thing. All of us have a special manner that we reserve for the microphone and children are no exception.

There are two ways of going after candid records: you can hide your Grundig and microphone completely, or bring them out in the open but pretend that you are not making a record. It all depends on circumstances.

If you decide to hide the tape recorder, the easiest way is to put it in another room or in a cupboard. This leaves you with only the microphone to think about, and the microphone is easy to conceal. The Grundig microphone extension leads allow you to have up to 45 ft. of lead from recorder to microphone, so you can place it on a shelf or even under the table. Don't put it inside an open box or an ornament or you'll get all sorts of queer sound effects. On the other hand, a fabric covering will not affect the sound, so you can stand the microphone on a window ledge with the curtain drawn, or fix it inside the shade of a standard lamp.

Plan the operation so that the victim's movements will be restricted to a particular area—*e.g.* make him sit in a special chair near the hidden microphone. Make a test beforehand to find the right setting for the recording level.

Set the speed control to the slowest speed and load the recorder with the largest reel of tape it will take. If you are using a cupboard in the same room, make sure that there is enough tape to last the visit: it is apt to be embarrassing if the tape runs out and a loud click comes from the automatic stop during a lull in the conversation.

You can either start the tape recorder just before your victim arrives or make an excuse to slip out and start it when you already have him on the spot. If the recorder is in a cupboard in the room you can arrange to go there to get drinks and switch on the recorder at the same time.

If you have a Grundig remote control you can leave the switch in a convenient position—say under the table if you are sitting around it—so that you can start and stop the tape when you like without attracting attention. The control is practically silent in operation but you will have to shut the Grundig in a cupboard or put it in another room because the slight noise it makes may give the game away.

Use interview technique to get the other person to do the talking and introduce as many incidental sounds as you can to give atmosphere to the record—e.g., clinking of glasses or teacups, the sound of a cocktail shaker, opening and closing doors and the like.

When you play back a candid record made in this way you will be surprised at the long silences. What you remember as a continuous conversation turns out to be a disjointed series of remarks interspersed with blanks. This is where the incidental noises help to give life to the record. And even with plenty of sound effects to fill the pauses, you will have to work hard to stop the conversation from drying up. Of course, you can always take out the blanks when you come to edit the tape, but a little quick thinking when you are making the record can save you a lot of tedious cutting out and joining up later. Best of all, use the remote control to avoid the blanks *before* they get on to the tape.

Groups of people

Pretty soon you will want to take your Grundig to a party or a committee meeting or some function where a number of people will be present. For this you will have to modify your technique because you can't expect your single micro-

70

phone to cope with people at different distances without having some voices too faint and some too loud.

If you have a Grundig mixer unit you can get over the difficulty by having several microphones and either selecting one or mixing them all according to who is speaking. If you

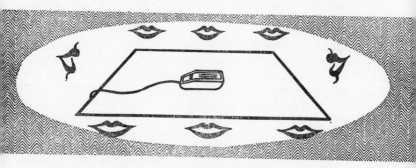

RECORDING GROUPS (CONFERENCES). *An omni-directional microphone (field shown white) is the right one for recording a round table conference. Make sure that all speakers are at roughly equal distances from the microphone.*

can't do this, you have to arrange for everybody to be at about the same distance from the microphone.

When everybody is sitting around a table—either having a meal or in a committee—just put the microphone in the centre of the table and set the recording level to give a reasonable signal without overloading the tape. Stand the microphone on a soft pad of foam rubber or cotton wool or it will emphasise the slight noises made when anything moves on the table—the rattle of cutlery and the thunder of dishes or rustling papers and tapping pencils.

If you are making records at a party or gathering where people are standing up and moving about, the easiest way is to take on a helper so that one of you can look after the recorder while the other does the interviewing. You and your assistant will have to work as a team. With a Grundig microphone extension lead the man with the microphone can wander around until he finds a victim—then signal that he wants to record; this is the cue for the assistant to start the tape and turn up the recording setting. The interviewer then goes into action with the interview technique, encouraging his prey to say something amusing or at any rate characteristic. Don't forget that the one who is conducting this interview should identify the speaker or get him to identify himself—otherwise, when you play the tape back later, you won't know who is speaking and you will lose a lot of the interest in the recording.

At the end of each interview the recording level control may be turned down to separate one interview from the next with a quiet period; or it may be even more effective to turn the level *up* to catch the general buzz of conversation and build up the party atmosphere. In any case always record a few seconds of atmosphere at the beginning and end of the tape.

If you have a Grundig mixer unit you can add a really professional polish to your recording by including a running commentary via a second microphone at the control position; here you can have an assistant doing the interviews and yourself the commentary—simply fading in the interview when you get the cue. You can also have a third microphone by the band, if there is one, and fade in the music to fill in any blanks. Or you can connect a gramophone pick-up to the mixer unit and add music from a disc to produce the same effect.

Recording live music

Once you know how to record voices you'll want to try your hand at taping music. Even if you only play a harmonica or strum a guitar, you'll be thrilled to hear what it sounds like to other people. If you are learning a musical instrument, of course, it will be a real help to have records of your playing. And if you and your friends who play instruments belong to a music group, your Grundig will be invaluable; after all, you can't make music and hear it properly at the same time—you only get the chance to study

72

music that you make yourself when you can record it and play it back.

But there's a big difference between recording speech and music. For a start, your voice uses only a small range of sound frequencies. A bass singer can get no lower than $1\frac{1}{2}$ octaves below middle C and a soprano can't get much higher than two octaves above it. So the whole range of the human voice covers no more than $3\frac{1}{2}$ octaves. And when you are talking normally, your voice does not use more than a single octave. But a piano uses over seven octaves and an organ eight, and for technical reasons you need to cover an even greater range if you want to capture the reality of orchestral music.

So you see that your microphone has a much tougher job to record music than speech—at least sixteen times tougher! But there is more to come. When you hear a record of somebody's voice, you are satisfied if it is clear and intelligible. It may be higher or lower or rougher or smoother than his natural voice, but you won't worry as long as you can understand what he says. But you know what the natural voices of a violin, clarinet or piano accordion sound like. and you expect them to sound exactly the same on your tape recorder.

In fact you will find that your Grundig will give you an astonishingly lifelike reproduction of music of all kinds and from all the instruments of the orchestra—as long as you give it a fair chance.

Easy musical instruments

You won't have much difficulty in recording single note instruments like the fiddle family, the wood winds, harmonica, recorder and such like. Just treat them exactly as if you were making speech records. First set the recording level by getting the player to run up the scale of the instrument from the lowest to the highest note. You need to do this because your recording amplifier is more sensitive to some parts of the scale than others. As a rule, if you adjust the level for the middle of the scale you will find that the

73

top notes will overload the tape and sound distorted. When you have found how much allowance to make for the extra sensitivity of your microphone to the high notes, remember it for future use: you can't always take around a musician to play test scales for you!

From this point you can go ahead with the normal recording procedure.

Difficult musical instruments

Percussion instruments like the xylophone, bells, and triangle—and the ones with strings that pluck, like the harp, guitar, mandolin and banjo—are a little more troublesome because of the way they make their musical notes.

When you strike a bell or pluck a tight string, the first thing you hear is the sharp knock or click that it takes to cause the vibration. This sound only lasts for a fraction of a second and it is followed by the true musical note. The musical note lasts for some time, getting fainter and fainter and finally dying away. The trouble with this sort of noise is that the first click or knock sounds very much louder to the microphone than the musical note that follows it. And if you set the recording level to suit the musical sound, the initial click is too loud and overloads the tape. The result is that the note sounds distorted. Instead of being a pleasing, pure note as it was in the original, it sounds broken and 'edgey'.

Don't think this is a problem that belongs only to your particular Grundig model—it is a constant headache to sound recording engineers everywhere. But in the Grundig everything possible has been done to get rid of this type of distortion for you. The only precaution you have to take is to watch the magic eye and keep the recording level just short of the point where it closes right up.

Remember that the part of the sound that causes the distortion may last such a very short time that the magic eye may close up too quickly for you to notice it. Everything may *look* all right to you but you may still be overloading the tape. So to be on the safe side, when recording instru-

74

ments of this type, start experimenting with the level control turned down to about half way below the point where overloading shows on the indicator.

When you make a record with the setting turned down in this way, all the other noises in the room become more noticeable. So you will have to take extra care to keep them down or they will spoil your record. And don't forget that sounds from outside the room all add to the background noises. You may not be conscious of them at the time because your thoughts are focused on the violin, guitar or harmonica you are recording; but when you play the tape back and all the sound is coming from the one point, you will get a painful reminder of the vacuum cleaner, the flushed cistern, the idling engine of the tradesman's van and all the other enemies of good recording. Of course you won't be able to stop all these noises, but you can choose a time when they are 'off the air', and you can shut a lot of them out by keeping all doors and windows closed during your recording session.

It is a great help if you can use a directional microphone —e.g., the GDM111—since this ignores a lot of unwanted noise and concentrates its attention on the subject alone. Here again, you can treat the solo instrument like a single voice. But don't have the microphone too close to the instrument or you will pick up a lot of incidental scraping and scrubbing.

The first time you make a recording of live music through the microphone, you will be so thrilled with the result that you will brush off all this stuff about noise and distortion as a lot of technical eyewash. But presently your ear will get more critical and you will be looking for some expert advice on how to improve the quality of your musical records. When that happens, you can always come back and read this chapter again. Nobody is going to say 'I told you so'!

Groups of instruments

In many ways recording music by groups of instruments is a different proposition from recording the voices of

75

groups of the same number of people. For a start, all the instruments have different kinds of voices; some are louder than others; and some affect the microphone more than others. Secondly, they may all be performing together or only one at a time. And finally, although there are some people who do most of their talking with their hands,

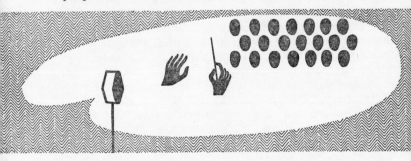

RECORDING PERFORMANCES. The cardioid type of field (shown white), as given by some dynamic microphones, records sounds from in front more strongly than from the sides or behind, so cuts out audience noises.

instruments usually want more elbow room than people, so they have to be allowed more space to work in.

When you want to record a committee meeting it is enough to set up the microphone in the centre of the table; but if you try to record a number of instruments that way the chances are that you will only hear the ones with the loudest voices.

The thing you have to aim at in recording anything from a piano and saxophone to a full orchestra is balance. If you can command several microphones and a mixing unit, you can place your microphones and balance the signals from them as required (p. 83). But with a single microphone you have to achieve balance by adjusting the relative positions of the music-makers and the microphone.

So long as you are not recording an actual performance, the easiest way to get a balanced record is by trial and error, using actual recordings to check your results and wiping them out afterwards. For a start, place the microphone near the weaker instruments and make a test recording. If you are dealing with a combination of first and second violins, cello and piano, for instance, start with the microphone about 3 ft. from the violins, the cello 6 ft. away and the piano at 10 ft. A test record will soon show you what changes to make to get the right balance so that no one instrument overpowers the rest.

Where you are simply recording a solo instrument or singer with accompaniment, it is easy to get the right balance either by asking the accompanist to play louder or softer or by adjusting the relative positions.

A singer accompanying himself on the guitar presents no special problems. Simply place the microphone 2 or 3 ft. away from him, run a test record, and decide whether you want the guitar played louder or softer.

Piano recordings

Piano and harp recordings call for special treatment because although there is only a single player, the sounds come from more than one point. On the piano the lowest and highest notes are about 5 ft. apart. So the problem is where to put the microphone to get the right balance between the low and the high notes.

When you get down to it you find that you can't lay down any hard and fast rules, because so much depends on the size and shape of the room, on the amount of furniture (and the number of people) in the room, on the wall covering, number of windows and whether the curtains are drawn or not. All these things play a part in the reflection of sound and its behaviour. And of course the type of microphone you use has an important bearing on the results, too.

Fortunately you can experiment as much as you like without having to worry about the amount of tape you are using. So spend some time trying the microphone at different distances from the piano, sometimes closer to the top notes and sometimes closer to the bottom ones. Try it with the top of the piano open or closed. Try opening and closing the doors and windows, with the curtains drawn and with them parted. Try the effect with and without pedalling. You may even find the best position for the microphone is actually *inside* the piano so long as it does not actually touch the frame; a piece of sponge rubber will insulate it.

If the room is fairly empty of furniture and without much in the way of curtains or carpeting, you may notice a peculiar hollow quality in the notes. This is caused by part of the sound travelling directly to the microphone and part reaching it after bouncing off the walls, ceiling and floor. In other words, echoes are getting mixed up with the true note.

There are several things you can do about echoes. You can try moving the microphone about the room (and the piano, too, if you are feeling energetic) until you find a position where the echoes are least troublesome. You can try drawing the curtains and opening the door or doors (this gets rid of the reflecting surfaces of the doors and windows). Most Grundig microphones hear sounds equally well from all directions; this is just what you want on most occasions, but a disadvantage when you are dealing with echoes. You can stop a lot of the echoes that strike the back of the microphone by fixing it in the open end of a big box or tea chest filled with cotton wool or soft rags. Or, better still, use a directional microphone like the GDM 111.

One final point when recording from the piano: don't forget that you are dealing with a percussion instrument and that you must keep the recording level on the low side to avoid distortion.

How the room affects your record

Everybody knows what an echo is—it is caused by sound waves bouncing off a hard surface and coming back to the point where they started from. When you face a wall or cliff at a distance and shout, the sound takes some time to reach the reflecting surface and travel back to you once more. So there is a gap between your shout and the echo, and you can recognise the echo for what it is. But if you are close to the reflecting surface—*e.g.*, inside a room—

78

the echo returns so quickly that it gets mixed up with the original sound and you don't realise that it is an echo. All you realise is that the sounds you record have a 'live' quality when there are bare reflecting surfaces around, and that

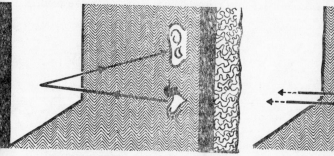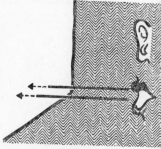

ROOM ACOUSTICS (A). When sound waves strike a wall with a hard, flat surface they bounce back and reach your ear as an echo a fraction of a second after you hear the direct sound. If the wall is covered with soft, porous material, it absorbs the sound waves and there is no echo.

they sound 'dead' if there are either no surfaces—like walls or ceilings—for the sound to bounce off or if the surfaces are soft and absorb the sound—like curtains or upholstered furniture.

When you shout inside a room or a hall, the noise goes on echoing around you for some time, getting fainter and fainter until you can hear it no more. This is known as reverberation. It is another factor that affects the quality of the sounds you hear. In a large public building the reverberations may take seconds to die away so that the sound of anyone talking, singing or playing a musical instrument is a mixture of original sounds and the reverberations of the sounds that went before.

In addition to all this, the quality of the sounds you record depends on the frequency—high notes are affected in one way, low notes in another—and the proportions of the room—a long narrow shape gives a different effect from a square room. It even depends on the weather—whether it is sharp and frosty, or dull and muggy—and the temperature. And it depends on where the sound source is in the room and where you have your listening point—*i.e.*, your ear or microphone of your Grundig.

Fortunately it costs you nothing to experiment until you find the right set-up for recording. You can make records

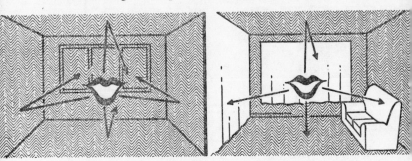

ROOM ACOUSTICS (B). When you speak in an unfurnished room your voice is reflected from walls, floor and ceiling. When there are curtains, carpets and soft furnishings present they absorb the sound waves and 'deaden' the sound of your voice.

with your microphone in different positions around the room; you can try it in different rooms; you can try it with doors and windows open or closed and with the curtains drawn or parted. All these changes will make corresponding changes in the quality of your recording. And each time you can erase the tape and try something different if you are not satisfied.

Make notes of each arrangement and the effect it produces, because you may not always want the same kind of quality. This is specially important when you come to record amateur plays or dramatic performances where a change of recording quality can help to create the impression of moving the action from one room to another or from indoors to outside.

You can also control the recording quality by your choice of microphone type and its position. An omni-directional microphone will give one character to the sound while a directional type will cut out a lot of the echoes and give it quite a different character.

If you find echoes troublesome—*e.g.*, when recording a percussion instrument like a piano—there are several things you can do to get rid of them. Try moving the microphone about the room (you can even try it *inside* the piano). Then try opening the doors and windows. If you can, employ a

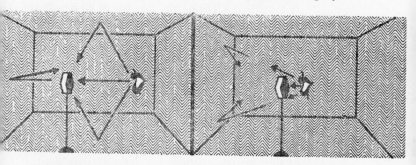

ROOM ACOUSTICS (C). *When you speak at a distance from the microphone, your voice may not record much louder than the echoes from walls and ceiling. When you speak closer, you increase the relative loudness of your voice and cut down the background noise.*

directional microphone and try altering its angle—side to side and up and down—as well as its position. Lower recording levels, with the instrument appropriately closer, also help to minimise echoes that otherwise would spoil recordings. Finally, try hanging lengths of cloth (blankets or carpet felt) from the picture rail at intervals against the longest walls.

On the other hand if your recording quality is dull and lifeless, try opening the curtains with the windows closed, roll up the carpet, and shift any large upholstered furniture into a corner. Increase the number of hard bare surfaces as much as you can in this way and you will add brilliance and crispness to your recordings.

Don't forget that the room acoustics have an effect on the playback quality of your records, too. Plenty of hard reflecting wall surface will tend to accentuate the high frequencies and exaggerate tape hiss—especially when you play back in the same room as the recording was made. A

F

lot will depend on where you place your Grundig and how you set the tone controls. You may find that certain low notes tend to 'boom' or that ornaments vibrate in sympathy with particular notes. It is always worthwhile attending to these things at the start so that you can enjoy the superb reproduction that your Grundig is capable of giving. Practically all the acoustic troubles affecting playback can be cured by placing your Grundig in the right position in the room.

Superimposing signals

Normally, you can only record one signal on the tape. You can record music from a gramophone pick-up, or speech from a microphone; but if you want to record the speech against the musical background you have to play the record through the gramophone and talk over the top of it so that both sounds go in through the one microphone and record as a composite signal. You have very little control of the balance of the two sound sources and you can't tell how the proportions of the mixture are working out until you finish and play the tape back.

In practice there are two ways of combining more than one independent signal on the tape—by superimposing and mixing. To mix signals the professional way you need a mixing unit, but even without one you can still get good results by superimposing recordings.

Superimposing means adding a second signal to a tape which you have already recorded. Normally, you can't do this because as soon as you switch to record, the erase head automatically comes into action and wipes off everything that was on the tape before. But on some Grundig models you can switch off the erase head with a button or key labelled Trick or Superimpose. This allows you to record a second signal on the tape without erasing the first.

To superimpose, you first play back the tape up to the point where you want to add the second signal. At this point you stop the tape, set the controls for recording from the required input and bring the erase cut-out switch into action. You can now start recording the second signal on top of the one already on the tape. As a rule you will have to make your second recording 'blind'—*i.e.*, without being able to listen to the results as you do it. This means that you will have to note the reading on the tape position indicator at the beginning and end of the section of recorded tape you intend to superimpose.

If your particular Grundig model has no provision for cutting out the erase head, there is still one way to do it, although not to be highly recommended. All you need to do is insert a smoothly folded piece of paper between the tape and erase head (the left-hand head); this will cut out most of the erasing effect—but be careful not to bend the spring on the pressure pad.

Mixing signals

Mixing means combining signals from several different sources and then recording the composite signal in a single operation. For this you use a Grundig mixer. A mixer has to perform some of the functions of a pre-amplifier—*i.e.*, it has to provide inputs to match the characteristics of a variety of signal sources—and it has to translate all the signals into terms that the amplifier can deal with. But whereas with most pre-amplifiers you can only switch one signal at a time through to the amplifier, with a mixer you can switch more than one; and you can control each signal separately to adjust the proportion of each one that goes into the signal you finally record. There are two principal types of mixer; one is simply a combination of variable resistances and the other incorporates amplifying valves. (This type has to be connected to the mains.) With the first type of mixer you can't boost a weak input, you can only level down the stronger inputs. In addition to the normal volume control on each input, this type of mixer has a separate pre-set level adjustment on channels intended for powerful signals—*e.g.*, from gramophone pick-ups or radio sets. When you want to mix these signals with the weaker signals that you get from a microphone, you first adjust the

83

pre-set control to reduce the strong input to the same level as the weak one. You can then treat all the signals on the same basis when you come to mix them with the normal controls.

When the mixer incorporates an amplifier you can boost weak signals as well as cut down strong ones, so you have a much greater power of control over the mixing operation. This type of mixer has a socket for a pair of monitoring

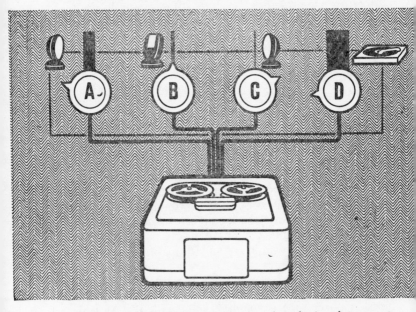

MIXER UNIT (PRINCIPLE). This four-channel mixer has separate controls for independently adjusting the signal level on each channel. As shown, four signals—strong A, medium B, faint C and very strong D —have been amplified or reduced by the related controls to mix at the same level and feed into the same tape recorder input.

headphones so that you can adjust the volume of each signal by ear and a magic eye recording level indicator so that you can keep a visual check on the total level of the combined signals.

There is a third type of mixer which, although it will amplify weak signals does not have to be connected to the mains. Instead of amplifying valves it uses transistors which draw their power from an internal battery. The The Grundig Stereo Mixer (p. 152) is of this type. It is powered by two dry cells and will handle both stereo channels from radio, discs or microphones.

You can use a mixer as a down-to-earth tool to simplify the business of recording from a number of separate sources— e.g., at conferences or for orchestral, choral and dramatic performances. Or you can use separate items creatively to build up an original sound picture. Either way you will find the mixer a useful addition to your equipment.

The basic operations of mixing are the same whatever mixing unit you are using. But the actual details of setting up before you start vary with the model.

CHAPTER 5

Tapes and Spools

MAGNETIC TAPE IS THE RAW MATERIAL FOR capturing all the interest and enjoyment you get out of your Grundig, so you ought to know the basic facts about it.

Briefly, it is a flexible ribbon $\frac{1}{4}$ in. wide, consisting of a base of polyester, between 1/1000 in.–1/2000 in. thick, and coated with a layer of magnetic iron oxide. The iron oxide is in the form of microscopically fine powder which is applied to the base in an absolutely uniform coating to maintain an even quality of reproduction. Rigid specifications are laid down for the magnetic and electrical characteristics of the tape. The brown colour of the coating on ordinary magnetic tape is due to the magnetic iron oxide which is nothing more than a pure form of rust.

There are four kinds of tape: standard, long play, double play and treble play.

Standard

Standard tape is tough, can be stored indefinitely, and has excellent playing qualities. Its playing time per 7 in. spool (1200 ft.) at $7\frac{1}{2}$ i.p.s. is almost 30 mins. per track—or 60 mins. for both tracks. The manufacture of this type of Grundig tape has been discontinued in favour of the long-play polyester type below, but you can still obtain other makes. If you choose another make of tape, bear in mind that your Grundig is pre-set to give best results only with recommended types.

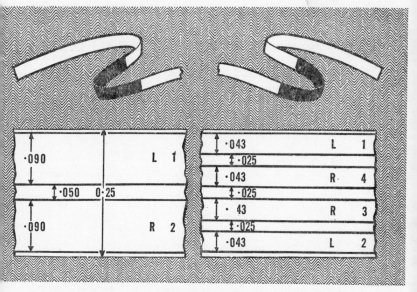

THE TAPE. The Magnetic tape used on Grundig machines is ¼-in. wide. It has a coloured leader tape at each end and most types also have a length of metal foil to operate the automatic end-of-tape stop (top). In 2-track recording (bottom left) you make two separate recordings, one on the upper half and one on the lower. In 4-track recording (bottom right) the tape carries four separate recordings: 1 and 2 are made by the upper head and 3 and 4 by the lower head. 2-track stereo tapes carry the R and L channels of a single recording; 4-track stereo tapes carry two sets of R and L channel recordings as shown.

PLAYING TIMES OF GRUNDIG STANDARD TAPES (EACH TRACK)

Spool Diameter (in.)	Length (ft.)	Tape Speed (i.p.s.) 1⅞	3¾	7½
7	1200	120 mins.	60 mins.	30 mins.
5¾	850	90 mins.	45 mins.	22½ mins.
3	175	18⅔ mins.	9⅓ mins.	4⅔ mins.

Long play

You can get 50 per cent more playing time from a given size of spool by using long play tape. Although thinner than

87

the standard type, it is practically free from stretch and can be used for accurately synchronised film sound. Long Play tape is no more expensive to run than Standard. The playing time per 7 in. spool (1800 ft.) at $7\frac{1}{2}$ i.p.s. is approximately 45 minutes per track—or 90 minutes for both tracks.

PLAYING TIMES OF GRUNDIG L.P. TAPES (EACH TRACK)

Spool Diameter (in.)	Length (ft.)	Tape Speed (i.p.s.)		
		$1\frac{7}{8}$	$3\frac{3}{4}$	$7\frac{1}{2}$
7	1800	180 mins.	90 mins.	45 mins.
$5\frac{3}{4}$	1200	120 mins.	60 mins.	30 mins.
3	250	$26\frac{2}{3}$ mins.	$13\frac{1}{3}$ mins.	$6\frac{2}{3}$ mins.

Double Play

The Grundig transistor portables and 4-track models use specially thin 'extra long play' tape, giving a greater tape length per spool. You can buy similar tape in all the regular spool sizes if you want to increase the playing time of your normal Grundig model. To find the playing time of a spool of double play tape, simply double the playing time of the equivalent spool of standard tape.

Remember that double play tape costs more *per foot* than long play tape, so there is no point in using it unless you really need the extra continuous playing time.

Triple Play

This is the thinnest—and most expensive—of all the Grundig tapes but it has the advantage of packing a long playing time on to a small spool—*e.g.*, a $4\frac{1}{2}$ in. spool of triple play tape holds the same length of tape as a 7 in. spool of standard tape.

Double and triple play tapes are particularly suitable for 4-track models because the extremely thin, flexible base makes it easier to hold the coating close against the record/playback head. Thicker tape tends to buckle and spring out of contact—a common cause of poor recording quality.

Pre-recorded tapes

Nowadays most of the leading gramophone recording companies issue their own pre-recorded tapes of everything from serious works to light music, dance music and jazz, and by this time there is a considerable list to choose from. If you do not want to go to the trouble of recording from the radio, you can buy your music ready recorded on to tapes in this way. Pre-recorded tape issues are reviewed in the periodical press along with the latest discs.

Leader tape

Leader tape is a non-magnetic, coloured plastic tape used at the beginning and end of a spool of magnetic tape. It is simply there to enable you to load the tape ready for use without winding on part of the magnetic tape to the take-up spool.

When you buy a spool of tape it will have about 3 ft. of leader tape at each end—often with one colour leading to track 1 and another leading to track 2. You can buy leader tape in a range of colours for use in making up your own tapes. You join it to the magnetic tape by splicing (p. 93) and you can insert it into a length of tape to mark the division between two separate recordings.

If you make a lot of different types of recording, you may find it convenient to use different coloured leader tapes for all the separate categories—*e.g.*, yellow for classical music, red for light music, white for talks, and so on.

Marker tape

This is an opaque coloured plastic tape which can be spliced or welded to magnetic or leader tape. It is stouter than leader tape and one side has a matt finish so that you can write on it with a pencil or ball-point pen. This type of tape is intended primarily for use in short lengths for labelling and marking recorded tapes, but it is sometimes used as a complete leader tape to do both jobs. Marker tape should be joined with its matt surface facing outwards on the spool so that you can read the information written on it.

Indexing your tapes

The tabular index printed on the inside or back of the spool carton is the simplest tape guide you can have for locating recordings quickly. It provides spaces for you to insert full details of every recording. In front of each recording you write the start and finish readings of the tape position indicator, or, if your machine has no indicator, write a description of the marker tape inserts you use instead (*e.g.*, red—or a written number on white tape). This enables you to go back to the exact starting point of any particular item on the tape.

The carton index is all you want if you only use one or two tapes. But if you are going to build up a library for business or pleasure, you will have to keep a complete index; otherwise you will have to search through the cartons every time you want to trace a recording. If you want to do the job properly, keep three separate types of index (they can all be in the same loose-leaf file). These are an alphabetical index, a spool index and a spare tape index.

The alphabetical index contains a list in alphabetical order of every item you record. A loose-leaf, ring-binder file with alphabetical separators is ideal for the job. You can use any system of classification that suits your subjects best, but don't have too many cross-references or you will be in trouble when you decide to erase a recording. Your entry should start with the abbreviated title, followed by the tape speed, the number of the spool, the track, and the tape position indicator reading (or marker tape key).

ALPHABETICAL INDEX

Title	Tape Speed i.p.s.	Spool No.	Track	Indicator Reading	
				From	To
Brahms *Symphony No. 4.* L.P.O. Sargent	$7\frac{1}{2}$	13	1	0	59

The alphabetical index won't help you if you want to know what follows or goes before any particular item. So you need a separate spool index which gives a list of everything on the spool. You can refer to the index on the spool carton for this purpose, of course, but it is much handier to keep a separate record in the same binder as your alphabetical index. Allow a page to each spool, divide it with a thick line across the middle, and enter track 1 particulars on the top half and track 2 details below the line. For a 4-track tape you will have to divide the page into four sections. It is always useful to take a quick look at the spool index after you have found what you want in the alphabetical index, because there may be another item on the same tape that you would like to play while you have it on the deck.

Enter the particulars in the spool index exactly as you do in the alphabetical index, but there is no need to fill in the track; this will be clear from the half of the page on which the entry appears. Example:

SPOOL INDEX

Spool No.	Title	Indicator Reading From	To
13	Brahms *Symphony No. 4*	0	59
	Scheherezade	61	79
	1812 Overture	83	107

There will generally be a short length of spare or blank tape at the end of every track.

Usually the amount will be too small for a serious record, but just what you want for short items and experiments. For this index you need only allow a single line for each track, and there is no need to enter any details of the recording. Simply tabulate the spool and track number, the tape position indicator reading (or marker tape key) at the end of the last recording, and an estimate of the playing time remaining at $7\frac{1}{2}$ or $3\frac{3}{4}$ i.p.s. It is better to stick to the same playing time throughout, and double or halve the figure if you want to play back at a different speed.

91

Spool No.	Track	Indicator Reading	Remaining Playing Time at $7\frac{1}{2}$ i.p.s. (mins.)
13	1	79	7
	2	89	$4\frac{1}{2}$
16	2	44	9

Tape spools

You buy magnetic tape ready-wound on to plastic spools. The spools have a hub with a central hole which has three equally spaced slots to fit the driving spindle either side up.

On the Grundig tape spools one of the spokes has a radial slit from the edge to the centre to allow quick attachment of

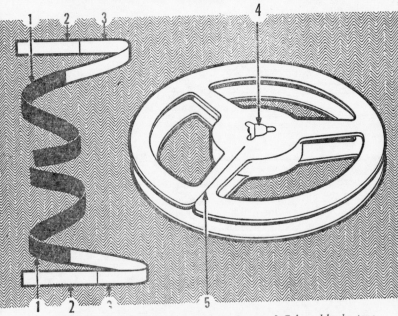

TAPE AND TAPE SPOOL. 1 Magnetic tape. 2 Coloured leader tape. 3 Automatic stop foils. 4 Hole recessed to fit spool holder spindle. 5 Tape slit for securing end of tape.

the end of the tape. The end of the tape is pulled along this slit until it reaches the hub, and then held by revolving the spool a few turns; the tape will then release automatically when it comes to the end without snapping or snarling up.

Grundig spools are made in three popular sizes—3 in., 5¾ in. and 7 in.—and the amount of tape they hold depends on the type of base—*e.g.*, a 7 in. spool holds 1200 ft. of standard tape or 1800 ft. of the long play kind (p. 86).

Automatic stop foils

All the Grundig tapes are supplied with a length of about 6 in. of plastic-backed metal foil, inserted next to the leader tape at each end. This strip makes an electrical contact which operates the automatic stop.

This is a useful safeguard if you aren't there to switch off at the end of the recording, and means that recordings can be safely made without supervision (*e.g.*, a long radio play).

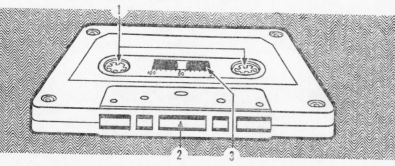

THE COMPACT CASSETTE. The tape is stored coated side out on 2 cores 1—1 which engage with the drive spindles. When playing or recording the record/play heads press against the coated side of the tape in the centre opening 2. A window and scale 3 let you see how much tape you have used.

Tape Cassette

The C100 battery portable model uses the DC-International tape cassette system in place of the normal tape spools. The cassettes are loaded with tape $\frac{1}{16}$ in. (4.6 mm) wide on ready-threaded spools and have only to be slipped

93

into position on the deck. The system allows two half-tracks to be recorded on the tape; when you come to the end of the first, you simply turn the cassette over and the tape will then run through once more using the second half-track. One

SPLICING THE TAPE. 1 Overlap the ends of tape, dull side down and cut through both ends at once. 2 Remove loose ends, butt remaining ends together and join with splicing tape. 3 Trim away surplus splicing tape, cutting slightly into magnetic tape.

cassette holds enough tape to run for approximately 2 hours at the standard playing speed of 2 i.p.s. (5.0 cm/sec).

Model C200, the latest cassette model uses Compact Cassettes, giving total playing times of 60, 90 and 120 i.p.s. It can be used for playing back standard pre-recorded Musicassettes. These are $\frac{1}{4}$ track stereo tapes, but because the right and left hand channels are recorded side by side they play back as a mono recording on the $\frac{1}{2}$ track C200.

Joining the tape

If your Grundig tape breaks you can join it up again as good as new; and you can cut parts out or insert them into a tape you have already recorded (p. 99). There are two common ways of joining tape; by splicing the ends with jointing tape, and by welding them with liquid solvent. (Cassette tapes cannot be repaired or edited in this way).

To make a spliced joint you first bring the ends of the break together and let them overlap by about an inch, dull side down. Hold the two sides firmly in place and cut across the double thickness of tape at an angle of about 45° with scissors or a razor blade. Now remove the top severed end, but don't let the two sides of the cut move in relation to each other. Press a piece of adhesive splicing tape over the cut and trim away the surplus from each side of it, also slicing off a thin strip of magnetic tape as you do so in order to leave

94

a suggestion of a 'waist' at the join (this is to prevent the join from sticking to the neighbouring turns on the spool). *Note:* Even a temporary join will give trouble if made with ordinary cellulose tape because the adhesive squeezes out

WELDING THE TAPE. 1 Overlap the ends of tape and cut through. 2 Turn tape dull side up and wet ⅛ in. of one end with jointing compound. 3 Wipe off compound (and magnetic coating).

at the sides under pressure and makes the tape stick to the next turns and to the heads and capstan drive.

The special adhesive jointing tape made for the job is coated with a harder adhesive with less tendency to ooze out at the sides; and the base is non-stretching.

To weld magnetic tape permanently you need a bottle of the special jointing compound sold for the purpose and a small camel-hair paint brush. First you overlap the ends of the tape as for splicing above, but dull side up. Cut across both thicknesses of tape at an angle of about 45° and remove the surplus end on top. Now wet about ¼ in. of the upper tape end with the jointing compound and immediately wipe it off; the magnetic oxide coating will wipe off, leaving the tape clear. Next wet the clear end of the tape once more and slide the other untreated end over the clear part so that they fit exactly. Finally, press the ends together and hold them for about half a minute.

You will find both splicing and welding easier and quicker if you use one of the special jointing blocks, or jigs and cutters, sold for joining magnetic tape. These accessories have a channel just wide enough to take the tape and a couple of pressure pads for holding it steady. Between the pressure pads the channel is crossed by 45° and 90° grooves which form trimming guides for a safety razor blade or (for extra particular people) a special non-magnetic cutter.

A jointing jig should always be fixed down on a solid surface as close to your Grundig as possible.

Looking after your tapes

Grundig tape is tough stuff and requires practically no

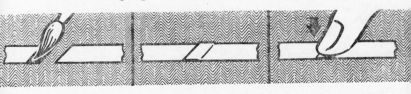

4 *Apply a second coat of compound to clear part of tape.* 5 *Lap other end of tape over compound while wet.* 6 *Hold ends of tape together for ¼ min. and the weld is then complete.*

maintenance. About the only things you need worry about are dust, heat and magnetic fields.

Dust need never be a problem if you simply make a habit of returning tape spools to their boxes when you have finished playing them. If you should leave a tape exposed in a dusty place so that it becomes thickly coated, blow or wipe the side of the spool clean before you play it, otherwise the dust will accumulate on the heads and affect the recording and reproducing quality.

Heat weakens the magnetic recording on the tape, and extreme cold tends to make the base brittle so that it breaks easily and does not run smoothly over the heads. So keep your tapes away from extremes of temperature.

It is always advisable after a tape has remained undisturbed on the spool for over six months to run it off on to another spool. This alters the relative position of the turns and prevents heavily recorded sections from printing through to the next layer of tape.

Magnetic fields set up by permanent magnets or by electrical equipment can weaken or completely erase the magnetic recording on a tape. The things to watch for particularly are the permanent magnets in the back of loudspeakers and in moving coil and ribbon microphones; also the small (but extremely powerful) permanent magnets sold as toys, electric motors (as fitted in hair-driers and other power-driven domestic equipment), chokes in electro-medical equipment and fluorescent lighting, and in fact anything where there is a coil of wire carrying an electric current—particularly if it has an iron core. None of these can affect the tape unless it is brought very close to it while running or while the current is switched on. But remember, once a recording has been weakened or erased by a magnetic field, nothing can put it back again.

Don't let tape get oily or greasy, because this will prevent the capstan from exerting a steady pull on it and cause wow or flutter. Eventually the oiliness will affect the capstan so that it will not even work properly with clean tape. Get rid of any oil on the tape, heads or capstan by wiping with a soft cloth moistened with methylated spirit.

For storage, stand your tapes in their boxes, upright on a shelf like books in a library; don't stack them in a pile one on top of the other. Filed like books you can then take out one tape without disturbing the rest. Number the tape cartons either by writing on the box itself or by sticking on the little numbered labels that you can buy at stationery shops. If you want to make things really easy for yourself, number all four edges of the box at the same height, and it won't matter which edge faces out. And even if you have spools of more than one size, keep them all in the same size boxes and they'll look neater and be easier to keep tidy.

When you buy a spool of magnetic tape it has a marker tape attached to each end. These tapes have a matt surface that you can write on in pencil or dry ink. If you have cut a tape into shorter lengths you can label the beginning of each track on the tape by splicing on a length of white or coloured marking tape.

If you simply want to give the complete spool of tape a distinctive marking, you can do it by splicing on a coloured leader at each end, using a colour code to indicate the subject (*e.g.*, natural history, music or family events). You can also insert lengths of marker tape in a spool to identify the start of different recordings; but remember that when you cut the tape you cut through all tracks, so only use this method when you don't care what happens on the other tracks. One way out of the problem is to stick a length of marker tape *on the back* of the recorded tape, without cutting through it. The double thickness of tape will bump as it passes through the sound channel, but it will do no harm. You can use different colours to mark the position of recordings on each track. And, of course, you will note these details in the index file so that you can pick out the section of the tape you want. This method is specially useful if your Grundig model is not fitted with a tape position indicator and you have to rely on other means to pin-point your recordings accurately.

Tape Speed and Sound Quality

Some Grundig models have only one tape speed, $3\frac{3}{4}$ i.p.s., some have two, $3\frac{3}{4}$ and $7\frac{1}{2}$, or $1\frac{7}{8}$ and $3\frac{3}{4}$, some have three, $1\frac{7}{8}$, $3\frac{3}{4}$ and $7\frac{1}{2}$ and the cassette model runs at 2 i.p.s.

The faster the tape runs, the better the recording quality but the more tape you use. This looks as though you should always use your highest speed if you want to get the best quality. But in fact you don't always need the highest quality your Grundig is capable of delivering. It all depends on what you want to record. The human voice has a very restricted frequency range and you can reproduce it quite well at $1\frac{7}{8}$ or $3\frac{3}{4}$ i.p.s. This is useful for making a little tape go a long way when you are dictating correspondence or if you want to tape a play from the radio. Then most Grundig models will give excellent reproduction of music at $3\frac{3}{4}$ i.p.s. In fact many Grundig owners regard anything faster than $3\frac{3}{4}$ i.p.s. as unnecessary and tape-consuming. However, if your model will give you $7\frac{1}{2}$ i.p.s. you will probably prefer to use it for the occasional musical work that you want to record at maximum possible quality. But make sure the quality of the signal is good enough; your higher recording speed will be wasted on an A.M. radio or a cheap record player.

CHAPTER 6

The Technique of Editing

FOR THE FIRST WEEK OR TWO, YOUR GRUNDIG will be a real magic box, and everything you record on it will sound perfect; just as the first snaps you took with your first camera were all masterpieces. Then, presently, you get more critical. A recording that sounded fine the first few times you played it back, begins to seem ragged and amateurish. One of the reasons is that it needs editing.

Editing does for a tape recording what careful enlarging does for a snap—it cuts out the unnecessary parts and adds emphasis to the really interesting things. And it can go further. With a photograph you can only take away things you don't want, but with a tape recording you can add other things that you *do* want, and then move all the separate items around to get them into better order.

There are two principal ways of editing: by cutting and splicing and by dubbing, with a second tape recorder.

Note: It is not possible to edit the tape in a cassette.

Cutting and splicing the tape

All the things you need for this have already been described in the chapter 'Tapes and Spools'. You can use the cutting and splicing method to remove a part that you don't want and join up the tape again, or to cut out the parts that you *do* want from a number of tapes, and join them up to make a separate recording. With a little practice, you can even cut a word out of a sentence.

For straightforward editing of this kind, first play back the tape you are going to edit, and make a note of the sound immediately before the beginning of the unwanted passage, and the sounds immediately before and after the end. Re-wind the tape and play it back again. As you approach the first cue word, get ready to press the stop button. If you have a slower tape speed available, switch over to it at this point; the pitch of the sound will drop, but with practice you will be able to recognise your cue and pinpoint it exactly.

As soon as you hear the cue, press the stop button. The place to make the cut is where the tape actually crosses the centre of the record/playback head. To get at it you will have to remove the sound channel cover. (You won't need to do this every time if you measure the length of tape between the centre of the head and the end of the sound channel; this will tell you how much tape to draw through the channel before you make a cut). The second cut—at the end of the unwanted passage—will probably have to be made by careful judgement; but if the length of it is sufficient, you can feed the fresh end through the capstan and play to the end of the unwanted section (there is no need to wind it on to the take-up spool; the capstan will pull it through.

If the piece of tape you want to cut out is too short to be worth saving, simply grip it between your finger and thumb and *pull* it through the sound channel with the control switched to playback. As soon as you hear the second cue sound, press the stop button and cut the tape as before. Now discard the cut out section, draw the end of the tape through the sound channel, splice it to the tape on the take-up spool, and wind the whole tape back on to the supply spool.

If you want to save the cut out length (*e.g.*, if it is a long piece), remove the take-up spool after you have made the first cut and put an empty spool in its place; then wind the next section of the tape on to the empty spool until you reach the second cue sound. Cut the tape and remove the temporary take-up spool with the unwanted passage on it. You can now replace the original take-up spool, join the ends of the tape and wind back on to the supply spool. This is the method you would use for changing the order of an item in a recording; once you have cut out the item as above, you can join it on to either end or insert it wherever you like in the main recording. This way you can record a spoken or musical introduction and add it to the original recording, or insert comments at any point in the tape.

In this type of editing you have to do everything possible to stop the joins from showing up. You must take care when you splice or weld the tape not to leave a gap without any oxide coating, because this will cause a click when you play back. And you mustn't join two tapes directly if there is a big change in volume or character between the two recordings. You can avoid jumps from this cause if you make a habit, when recording, of starting and finishing every item with the recording level control turned right down. When you join items recorded in this way, it does not matter how different they are; they will always join up smoothly. Your ear won't jib at the fade out/fade in effect as it would at a sudden jerk from, say, a roar of applause to an opening chord of music.

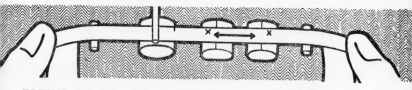

EDITING BY SEE-SAWING. To pin-point the spot you want— e.g., the beginning or end of a word—switch to playback and draw the tape backwards and forwards over the record/playback head while you listen for the sound through the speaker or headphones.

But you can save yourself a lot of cutting and piecing together by leaving a few seconds of blank tape before and after every item. This will give you space to add any spoken comments straight on to the tape—instead of recording them separately and then fitting in the odd piece of tape later with a joint at each end.

Precision editing

The above method is all you need for dealing with whole sentences with well defined breaks, but if you want to cut

101

out a word or part of a word there is an easy way of editing that you can use for pin-pointing even a syllable.

First use the above method to find the rough position of the sound you want to cut out. Now remove the sound channel cover and set the main control to playback, but do not start the tape: on some models you simply press the playback key without pressing the start key; on others, where the tape automatically starts when you switch to playback, you use the temporary stop control. Now take hold of the tape at each end of the sound channel and 'see-saw' it to and fro over the record/playback head. This will play back the section of tape well enough for you to identify the point you want, and you can then narrow down your see-sawing until the spot lies exactly on the centre of the head. You can mark the position with a Chinagraph pencil and then pin-point the other end of the unwanted section in the same way.

If the section is very short—*e.g.*, a syllable—there is no need to cut the tape; you can erase the signal. There are

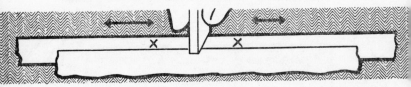

PRECISION ERASING. *You can erase a word (or part of a word) by see-sawing the tape over the erase head gap with the control set to record. Or you can erase it (after pin-pointing and marking) by drawing a permanent magnet along the tape.*

two ways of doing this—with the erase head on your Grundig and with a permanent magnet.

First, mark the start and finish of the section you want to erase. Now switch to record and draw the marked section of tape over the erase head. There is no need to see-saw it; a single passage over the head is enough so long as you keep the tape firmly in contact with it.

The other method is to draw one pole of a permanent magnet over the marked section of tape. One of the small, powerful toy magnets sold for a shilling or so is excellent for the job. If you do it this way you must be careful to keep the magnet away from the rest of the tape. To be on the safe side start by practising on a spare length of recorded tape. By holding the magnet a short distance away from the tape you can even fade a section in and out.

Remember when you cut out a section of tape you can always splice it back again, but when you erase it there is no way of restoring the missing recording.

Continuous loops

If you join the ends of a length of tape to make a continuous loop you can run it through the sound head of your Grundig and play it back non-stop. This is one of the most useful tricks in editing. You can use it for continuous background effects—*e.g.*, crowd noises, rain, wind, waves, traffic and so on. You can use loops as sound effects for live performances or for dubbing on to feature tapes from a second recorder.

When you make a tape loop take extra care to make a

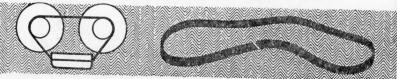

ENDLESS LOOPS (A). *For producing continuous sounds you can join a section of recorded tape end to end to form an endless loop which will run non-stop through the sound channel. A short loop can be passed around a pair of empty spools.*

smooth joint because it will have to run through the sound channel again and again. And the start and finish of the section you choose for the loop must join up without a noticeable change in volume or a break in the actual sound— *i.e.*, if you are making a loop from a recording of applause, you must cut in and out at the same signal level and you must not cut the tape in the middle of a shout or a whistle. If you are making a loop of rhythmic sounds like the sound of a railway engine or a big drum make sure there is the same interval between sounds on the piece of tape with the join in it as there is on the rest of the loop. This means pulling the tape past the playback head repeatedly until you have carefully located the required spot—then marking it so that you can cut the tape correctly to at least $\frac{1}{8}$ in. at each end of the loop.

When you play the tape loop you must load it on to the recorder so that it runs freely. Where the loop is short, you can simply thread it around a couple of empty spools fitted to the spool holders in the normal way. If you want to use a longer loop, you can pass it around a jam-jar placed on a box level with the top of the deck. If it is a long loop that you only want to use for a short time, you can even let it run around a smooth pencil held in your hand.

Endless loops are used for making musique concrète— *i.e.*, for building creative sequences of sound. These start with basic sounds recorded on tape and then run through the sound channel at different speeds or manipulated in other ways to give new sounds. The new sounds can then be combined by editing to produce a 'composition'.

ENDLESS LOOPS (B), *If you want to play back a loop that is too long to go round the spools, you can improvise a pulley from a glass jar supported or weighted down at a suitable distance from the tape deck.*

Composite recordings

By combining these editing techniques you can get up to all kinds of tape tricks. You can, for example, carry out an interview in which your victim answers a succession of prepared questions. Later, on the editing bench, you take out the questions and substitute others which put the answers in a totally different light.

For example, the original question might be: 'Do you prefer powder or paste for cleaning your teeth?' with the answer 'Oh, I wouldn't dream of using anything but Dento toothpaste.' Later you substitute the question: 'Do you prefer milk or cream in your coffee?' To anyone who recognises the voices of the parties to the interview, the result is always good for a laugh.

When you combine recordings made at different times in this way you should always make sure that the various parts are recorded under similar conditions—*i.e.*, either in the same room or in one with the same sort of acoustics and at the same distance from the same type of microphone, otherwise the joins will show.

Now see if you can build up a conversation between two people who have never met, make yourself or a friend take part in a concert with famous personalities; or have a conversation with a popular singer. This is the way to start having real fun with your Grundig.

Dubbing from another tape

The simplest use of dubbing (p. 58) is to transfer or rearrange the sequence of recordings. If you have a complete tape containing a mixture of dance music, party pieces and interviews, you may want to bring all the dance music together into one shorter tape, all the party pieces together in another, and so on. By splicing you can do this—but it's a tedious job; with a second tape recorder it's easy. First you record the opening piece, then stop both recorders and wind the original tape forwards or backwards to the next piece you want to add; and so on, recording each item separately in the sequence you decide. Always allow a slight gap on the new tape between recordings and start and finish with the volume turned down.

Dubbing is also useful for building up composite recordings and sound effects. Here you use the second tape recorder merely as an additional source of sound—just like a gramophone or radio. You then either mix or superimpose the output from it with whatever else you wish to employ. The method is no different from straightforward mixing or superimposing except that your whole tape library is now available as a source of raw sound material.

CHAPTER 7

The World of the Grundig

YOUR GRUNDIG IS PROBABLY YOUR MOST versatile possession. There is literally no end to the entertaining and useful things it can do for you. It can give you and your friends endless fun, it can add to the interest of your hobby or your studies, it can help you in your business or profession.

Many people never get further than the fun. They may think they are getting their money's worth even so, but they are only scratching the surface of its possibilities. You can use a pencil to leave a message for the milkman, write a poem or draw a picture; and in the same way you can use your Grundig as a toy, a tool or a means of creative expression. Up to a point you can learn from books or from other users, but finally a great deal depends on using your own imagination. People are continually discovering new uses for their Grundig equipment, but so far only the fringe of its possibilities has been explored. So use the ideas that follow as a beginning and go on from there; there's an exciting world ahead of you, waiting to be opened up—the world of the Grundig. . . .

Advertising

In the commercial world, a Grundig can help to advertise and sell products. Advertising agencies use Grundigs for a variety of jobs— *e.g.*, for recording radio and television 'commercial' ideas; and for recording interviews with the public in the course of market research and motivation studies (*i.e.*, finding out what makes people want to buy things). Creative teams in some agencies have 'brain-storming' sessions in which the members sit around a Grundig and think up ideas for advertising a product. The whole object of the session is to get out a lot of ideas, serious, humorous or completely farcical and to sort them out at leisure later. This system is claimed to lead to less inhibited thinking and is said to produce enough bright ideas to be well worth the trouble.

If you run a shop of any sort you can make your Grundig work for you to increase sales in a number of ways. One of the most effective ideas is to put on a tape of bird songs and play it through an extension speaker in the doorway of the shop (so long as there are no by-laws against it). Or you can tape a sales talk about a product on display and switch it on at suitable moments. You can even arrange matters so that the customers can press a button and start the tape playing back.

If you are a manufacturer you can be sure that your sales message is getting over if you tape it so that your representative can play it back to the buyer when he calls on him. If you can tie up your talk

106

to a series of colour transparencies so that your salesman can show your product in a portable viewer, so much the better.

You can also collect a lot of valuable consumer reaction data at exhibitions where your product is on display. All that is necessary is to install the microphone at a point where it will pick up the comments of the public without being seen. You get more unbiased information from a candid recording made in this way than by questioning people.

Album recordings

Keeping a tape album or scrap book can be one of the most rewarding uses of your Grundig, but you have to be systematic about it or the thing will just fizzle out like your entries in the diary you start every New Year's Day.

Set aside a spool of tape to record your album material and give it a distinctive, coloured leader tape so that you don't get it mixed up with other tapes. Before any event you think might be interesting to recall in later years, slip this spool on to your Grundig and record just enough of the event to give you the essentials. At the end of the recording, insert a slip of paper under the tape where it runs from the supply spool and wind back the tape. The next time you want to add an item you simply replace the album spool, fast wind the tape forward until the slip of paper drops out, and start recording at that point. This saves you the trouble of noting the tape position indicator reading each time. Keep a notebook with dates, places and names opposite each recorded item.

What you record on this tape is just the material for your collection. This material will have to be edited (p. 99) before it will be worth listening to later. If you can borrow another tape recorder occasionally, it is always better to dub the edited version on to a second tape; this way you can fill both tracks before you start editing. If you have to edit by cutting and splicing the tape, you will have to do it before starting on track 2 or you will cut into the recording on the second track. You want to be able to use up track 2 for a recording that does not require editing.

By waiting until you have a full spool before doing your editing you will give the items time to fall into perspective and you will have a better idea of what to cut out. You will need to record your own introduction—and possibly pay-off—to each item, so if you have to edit by cut-and-splice, leave a long enough gap between the items to let you do this without piecing in an extra length of tape. Keep your comments short, or the effect will be tedious and boring, and don't add humorous touches of your own; just stick to the facts and keep the thing as impersonal as a B.B.C. news bulletin. The interest is in the items themselves—not in what you might have to say about them. If they can't stand on their own without your help, they shouldn't be there anyway.

Your choice of subjects for your tape album will depend on your own tastes and the sort of world you live in, but you should keep the subjects as varied as possible; tape only real high spots, and edit ruthlessly. To start you off, here are a few suggestions:

107

Visit of your church carol singers.

Family weddings—church ceremony if possible, wedding breakfast certainly.

Twenty-first birthdays.

Family get-togethers with candid recordings of newcomers on arrival speeches and junketing generally.

Guy Fawkes night.

Voices of all family pets—dog, cat, budgerigar, etc.

Sample of baby's first words.

Family accomplishments—musical or dramatic.

Candid records of voices of family friends.

Excerpts from broadcasts of events of national importance—*e.g.*, speech by the Queen, final of the Boat Race and other classic sporting events, the last few moments of the final night of the Proms, and so on.

When you have finished your editing, you should either note the tape position indicator reading of each item or, better still, mark and number the start of each section with a piece of marker tape on the back of the tape. Enter the numbers and titles in your record index so that you can go straight to any item you want to play back.

There is no reason why you should stick to a general scrap book. According to your interests, you can build up all sorts of specialised collections, characteristic 'scraps' recorded either from radio or borrowed records or even live: popular comedians, singers, solo instrumentalists, politicians, famous people, and the like. Your Grundig lets you pick out the truly characteristic sounds and string them together in a record that grows in interest and value with each item you add to it. You can get the same sort of thrill out of this sort of collection as you can from any other branch of collecting, with the difference that you don't have to pay for the items.

Remember that your tape album will only come into its own after some years have passed. Listening to scrap-book programmes on the radio will give you a good idea of the kind of things that hold or increase their interest with the passage of time.

All in a day

You can have a lot of fun simply taping the sounds that go with your normal day's activities.

Start with the loud and metallic ringing of the alarm clock (or if you want to open up with a laugh, you can start with a snore and have the alarm cut it short). Then go on to the running of the bath water, shaving noises—*e.g.*, stropping your razor or the buzz of an electric shaver—the opening of the early morning radio news, slamming the front door, starting the car, and finally fade out on the sound of the engine dying away down the road. You can follow this with the barking of your dog to welcome you back in the evening, the sound of the cocktail shaker, the dinner gong, washing up, the start and finish of your favourite television feature, yawning, winding the clock—and last of all the distant striking of your church clock at midnight. (Season with the hooting of an owl if you have one in the neighbourhood.)

String all these sounds together and you will have a little five-minute feature that will bear constant repetition and that you can exchange with your tape friends. There will be no need for you to speak all through the record; the sounds themselves will tell the story. But if you want to add a professional touch of continuity to the whole thing, intersperse the separate items with a suitable piece of music of the *Perpetuum Mobile* type, fading out the music and fading in each sound in turn.

There is no need to stop when you have recorded your own day. The incidental sounds are different for every member of the family—mother, the children and even the household pets—and each one will make an interesting record.

Recordings of this kind can have a special sentimental value. Friends or relations overseas will always appreciate a tape which brings them a reminder of home. And even the sounds you associate with your holiday at the seaside can bring it back again for you later.

Bedtime stories

If you have the sort of children who like to be read—or sung—to sleep, your Grundig is ready to take the job off your hands whenever necessary. Simply record a collection of bedtime stories when you can spare the time—read them from a book or make them up or get a favourite uncle or aunt to do it—and you will be able to leave your Grundig to deputise for you when the need arises. To be on the safe side you should have your Grundig where you can keep an eye on it and run a lead to an extension speaker in the children's bedroom.

If you want to do the job properly, fix the Grundig microphone over the bed and run a second extension lead back to your microphone input. When the story comes to an end, set the main control to record or to straight through amplifier, depending on the model, and switch to microphone. You will now be able to hear the result of the bedtime story or lullaby. No matter whether you hear measured breathing or cries of 'Encore', you will be able to take the appropriate action without leaving your comfortable chair.

And don't forget, if your stock of suitable stories or songs runs out, you can always supplement it by recordings from the radio. And if you have a tape friend, you can always exchange bedtime tapes to introduce further variety. This way you each get twice as much work out of your tapes before you need to change the programme.

Birds

Recordings of bird songs are immensely popular—whether you know the difference between a blackbird and a thrush or not. And they are quite easy to make, wherever you live.

If you have a garden at the back no bigger than a pocket handkerchief there will be bird songs for your Grundig to record—particularly if you regularly provide food (out of reach of the cat). Practically every one of the common garden birds has a song worth recording—blackbird, thrush (missel and song), robin, chaffinch, wren and great tit. You can usually count on getting all these within recording range of a microphone placed on your window sill.

The time to hear all the singers hard at it is, of course, just after daybreak, and you should certainly make a recording of the dawn chorus for your collection. But listening to the dawn chorus is like listening to a room full of people all talking together; selected solo performances are much more interesting. To get good solos you need to study the habits of you bird population. You will soon find certain birds singing on their own at set times of the day, in their particular season. For example, the missel thrush will give long recitals at dawn and dusk in December and January, the blackbird will often lead the dawn chorus and go on singing half an hour after the other birds have fallen silent again. One way or the other there is always a time of day at a particular season of the year when you can be sure of getting a particular songster on its own. And most birds have a favourite singing perch.

Once you have established the right time and place, get the microphone as close as you can to your subject. If you are lucky enough to have the bird singing close to the house, you can do your recording with the microphone by the window, otherwise you can add a microphone extension lead up to 15 yards from the house and still have your Grundig indoors. For greater distances you will have to take your Grundig out of doors and run an extension cable from the nearest electricity point. The important thing is to place the microphone close enough to get a full-strength signal on to the tape; if the microphone is too far away and you have to turn up the signal level control beyond about halfway, you will get a lot of unwanted background noise. At the same time, remember that some birds deliver a surprising volume of sound; so you must be careful to avoid overloading and consequent distortion, which is particularly noticeable in this type of record.

If you want to go in for recording wild birds, you will need one of the battery operated models so that you can operate away from the electricity supply. You can't work too far away from a motor road with a converter because this and the battery will add up to quite a weight. The most convenient set-up if you have to rely on this type of supply is to have all the equipment in the boot of a car and run a long microphone extension lead (p. 154) to the recording location.

Wild birds are a very different proposition from the garden varieties. They are more shy, and it is difficult to pin their performances down to a time and place. This is a job for the specialist, and unless you have made a close study of the subject you won't get very far. A good book on bird photography will help you because there is a lot in common between taping and photographing wild birds—e.g., in the use of hides to conceal yourself and equipment. Otherwise what has been said above about recording garden birds holds good—i.e., in the need for getting close and recording at full level.

Children

If there are children in your house you have a subject that can keep your Grundig busy twenty-four hours a day.

You can tape the whole story of speech development, help the youngster to correct bad speaking habits (p. 123), add the first proud

110

'party pieces' to the family album, help to develop musical and dramatic appreciation, and promote imagination and creative thinking at the age when the child is beginning to take notice.

If you can compile a complete tape record to illustrate the gradual development of the baby's speech from the first articulate sounds to the first words and phrases, you will possess something that will never lose its interest. But don't make the mistake of drawing it out to the point of boredom. No example should last more than four or five seconds, and each one should be separated by an equal period of time—say, a week or a month—to give a measure of the rate of development. Make your spoken introduction to the series as short and factual as the label on a museum specimen. Before each weekly or monthly recording in the series simply interpolate 'second month . . .', 'third month . . .', and so on. Record plenty of tape each time, but edit it down to a few seconds of the best from each section.

If you want to make candid recordings of children, there is no need to hide the recorder and microphone. Put everything on the table or on the floor and pretend to be busy with cleaning or adjustments. The children will start by being interested, but as soon as they find nothing is going to happen they will get bored and start playing by themselves—especially if you have taken the trouble to provide a counter-attraction in the form of an indoor game.

Once the youngsters are safely shut up in a world of their own, you can set the microphone down somewhere near them and get to work. So long as you get your attention glued to the signal level indicator (which you'll need to do anyway) and remember to tell them to be quiet from time to time, they won't suspect a thing.

If the children are likely to be running about from one room to another you will have to turn down the volume and stop the tape when they get out of range of the microphone, unless you have extra microphones and a mixer unit. With a mixer unit you can have three or even more microphones distributed over the target area and switch over from one to the other as the action changes ground.

A word of warning. Don't make the mistake of letting the youngsters hear the record, or you'll never be able to work the trick again.

Cine films with sound

If you are an amateur cine enthusiast, a little help from your Grundig will make all the difference to your film shows.

It is the easiest thing in the world to provide a suitable musical background to a cine film. There is no need to synchronise it accurately with the picture, and it is only necessary to arrange for the mood of the music to correspond roughly to that of the film. You can start your Grundig just before the film, and once the projector is running turn up the volume control to fade in the music.

If you want to synchronise sound effects you will have to use an accurate method of starting and keeping the projector and tape in step. There are various synchronising attachments available which will do the job. Usually you have to set up the synchronising equipment alongside your Grundig and run the tape around a pulley on the synchroniser either before or after it leaves the sound channel; this in turn controls the speed of the projector to match. In some types of synchroniser you watch a stroboscopic disc driven by the tape and adjust the speed of the projector to hold the stroboscope steady.

With these attachments, so long as you start your Grundig to coincide with a cued frame on the film, the sound and picture will stay in synchronism. But no equipment of this type is suitable for really accurate synchronisation (such as lip movements with speech); for this you have to put the sound recording on to a magnetic strip on the actual film.

Dance music

By taping suitable numbers from your radio or gramophone records you can use your Grundig to provide music for informal dances or for practising dance routines. In this way you can collect a series of numbers to give you the same rhythm from start to finish—if you simply want to practise—or you can make up a varied programme to last a whole evening. If you want to give a realistic touch you can dub applause in between the numbers, and add encores in the traditional manner. By recording at a speed of $3\frac{3}{4}$ i.p.s. you will get all the quality you want plus the economy of recording three hours of dance music on a single 7 in. spool of tape.

Remember that you must not play dance music recorded in this way for anything except your own private use or for entertaining your own friends in a private house, otherwise you will be infringing the Copyright Act.

Dictation

While your Grundig is not intended to take the place of your regular office dictation equipment, it can do the job well (and in some respects very much better). For this purpose the foot-operated remote control (p. 152) available on some models is a big help. It allows you to pause while you are thinking of the next word and leaves your hands free for handling books or papers. And it allows your typist to transcribe at a speed to suit herself and to play passages over again by changing the pressure of her foot on the pedal.

If you are the type that prefers to dictate while walking about the room, the sensitive microphone of your Grundig will pick up your voice wherever you are. And if your typist does not like to listen to the record on the stethoscope earphones (p. 153), she can play back through the loudspeaker and adjust the volume to her liking.

Discs from tape

When you record at a speed of $7\frac{1}{2}$ i.p.s. the cost of the tape works out at rather more than you would pay for a long-playing gramophone disc of the same playing time. Against this you can set the advantages that your tape recording will never deteriorate, however often you play it, and you can use it for recording another piece of music when you are tired of the first recording. You can do this thousands of times if you like and the tape will still be as good as new at the end.

But if you ever want to put one of your tape recordings on to a disc so that you can play it on an ordinary gramophone, there are a number of specialist firms who will do it for you. The cost is about equal to that of the equivalent commercial disc—either long playing or 78 r.p.m.

If you wish to have discs made in this way, the tape should be properly edited beforehand so that it will run for the same time as the disc size you need. Don't forget, too, that if you are having both sides of the disc cut, your original recording must have a suitable break in it at a point roughly equivalent to the change-over of the disc from side one to side two.

Ideas—with a twist

Practically every day someone discovers a new application for the Grundig. Once you get out of the rut there is no end to the uses that will occur to you.

Here are a few ideas that other Grundig users have thought up; they will help you to get new ideas too.

An amateur pianist records difficult pieces at $3\frac{3}{4}$ i.p.s., playing the items an octave lower than written; then he plays them back on the recorder at $7\frac{1}{2}$ i.p.s. The result is a brilliant rendering at a speed that would be impossible for the most gifted professional.

A manufacturer wants to telephone to Australia and tapes the information at the slowest speed and transmits it at the highest speed. At the other end it is recorded at high speed and played back at slow speed. In this way the manufacturer gets twice as much telephone time for his money. (But he must first consult the telephone authorities).

A motor repairer makes records of all abnormal engine noises and their causes. This helps his staff to diagnose engine trouble by comparing the customer's noise with the taped specimens.

A watchmaker checks the movement of a repaired watch by resting it on the microphone and listening to the amplified sound of the mechanism. Slight irregularities can be detected and the constancy of ticking easily checked.

A dog trainer, in the course of teaching a 'pupil' to stay quiet, leaves it indoors with the Grundig switched on. When he returns after an hour or so he plays back the tape to see if the dog has barked or whined while he was away.

A model maker uses pre-arranged recordings on tape to provide electrical signals (from the outlet socket) to control working models. The signals on tape can be of any duration or frequency to suit his purpose so that he can get the model to go through any sequence of operations automatically. (The Grundig Sonodia Slide Changer (p. 156) can be adapted to carry out any switching sequence for this type of model. It enables you to record impulses on the tape which, when played back, close the switch contacts in the Sanodia and so control any external circuit connected to them.)

A sound engineer joins a short length of recorded tape into an endless loop so that he can produce a continuous background sound (*e.g.*, the wash of the sea) at will.

If you have a 35-mm. camera you can make up your colour transparencies into slides or film strips and project them to the accompaniment of a sound track provided by your Grundig. This is a splendid way of putting over a holiday or giving continuity to a number of separate shots showing different aspects of the same subject.

There is no need to make a continuous film strip if your projector will take transparencies in a magazine. You simply arrange them in the magazine in the correct order and change them at the appropriate points in the recording.

Start by going through your pictures and arranging them in a smooth sequence. Then write a script around the picture sequence and mark in it the change-over points for each slide. Keep your comments short and don't waste time describing anything that is already clear from the picture.

Now record your script and at each planned change of picture give a couple of taps with a pencil on the table and wait a second or two before going on to the next part. When you play the tape back, this short pause will give you time to change the transparency (or wind on the film strip to the next frame).

If you have a Grundig mixer unit, you can show the pictures to a background of music, and introduce sound effects—*e.g.*, running water, aeroplane engine, bird song—to give continuity and atmosphere.

When you are showing a collection of individual transparencies you can arrange them to suit the interests of your audience and record a special tape beforehand—even if you are not likely to want it on more than one special occasion. This way you get through all the business of deciding the projection sequence and what you are going to say in advance. When your audience arrives you can enjoy the show with them. And if there are any youngsters in the house you can let them have the thrill of running the whole show, knowing that nothing can go wrong. If you really want to do the thing in style, the Grundig Sonodia Slide Change Adaptor (p. 156) will automate the whole process for you and go through the whole show—words and music—without any assistance whatsoever.

This is a device by which the tape recorder can be made to operate the slide changing mechanism of a transparency projector. It consists of a metal box with a sound channel on the top. The box is fixed by an adjustable clamp to a foot plate on which the tape recorder stands, and the channel can be levelled with the tape deck of any Grundig model. Two flexible leads connect the adaptor to the main electricity supply and to the remote control socket of the automatic slide projector. The tape is looped from the sound channel of the tape recorder, around the sound head of the adaptor and back on to the take-up spool.

The recording button on the adaptor puts a magnetic pulse on the tape, and when this pulse passes through the sound channel during playback, it operates the slide changing mechanisms. By this means it is possible to synchronise the slide changing sequence with music or a commentary recorded on the opposite track of the tape. It is also possible to make the adaptor operate any electro-mechanical equipment where the movement is initiated by closing a pair of contacts.

114

To use the adaptor for synchronising a set of slides with a commentary and sound effects, you first arrange the slides in order and write a script around them. Next you record the script on tape, together with any sound effects or background music you want to include. Next set up the Sonodia with the tape looped around the head and play back the record you have just made. As you come to each point in the script where you want to change the slide, you press the record button on the Sonodia adaptor. This records a magnetic pulse on the lower 1/4 track of the tape.

Now you wind back the tape, connect the adaptor lead to the slide projector remote control socket and load the slides into the magazine. When you play back the tape, your commentary and sound effects will reproduce normally and every time a magnetic pulse on the bottom track passes over the sound head of the adaptor, the slide will change automatically. If you have timed the pulses correctly, the slide on the screen will always synchronise with the commentary you have recorded about it on the tape.

Manufacturers nowadays make great use of the tape and film strip combinations for demonstrating and selling their products. A representative with a Grundig and a film strip projector or viewer can deliver a sales talk recorded by an expert, with colour illustrations showing the product in the course of manufacture, completely assembled, and finally in use. And you can use the same technique for putting over ideas, stories and propaganda. In many cases it is both the most effective and the cheapest way of doing the job.

Medicine

There are very many uses for tape recording in medical practice, in hospitals and clinics, and in the dentist's surgery. While many of these applications call for specialised recording equipment, a number of them come within the scope of an ordinary tape recorder.

Many doctors keep a Grundig handy in their surgery or consulting rooms so that they can make their own records of abnormalities in the patient's breathing, speech, heart functioning and the like. Some of them find it useful to play back these recordings to the patient to demonstrate an improvement which could not be shown in any other way.

The value of a tape recorder in psychiatry and in recording actual consultations has been recognised for some time.

In this and similar applications it is often best to keep the tape recorder out of the way and operate it with the Grundig remote control (p. 152).

Messages

There are all sorts of ways of making your Grundig earn its keep by taking messages. The most obvious one, of course, is to send a message to a distant friend who also owns a tape recorder. For this you use a special 3 in. message spool that holds 200 ft. of standard tape or 300 ft. of long-play tape and runs for 10 mins. (standard) and 16 mins. (long-play) per track. This type of spool is usually supplied in a postal box with a message form inside to carry written particulars of sender, subject, date, recording speed, etc.

You can record your message on both tracks and leave the other person to record a full-length reply in its place; or you can use one track and leave the second track clear for a reply, so that you have a record of the whole story. If you reckon on talking at about 120 words per minute you won't be far out. There is no point in using a fast speed for speech, so you should record at $3\frac{3}{4}$ i.p.s.—or $1\frac{7}{8}$ i.p.s. if your Grundig will work at that speed. Assuming that you record at $3\frac{3}{4}$ i.p.s., you will have time for about 1200 words on one track of a standard tape and 2,000 on a long-play.

While recording messages you need to keep your finger ready on the temporary stop control so that you don't waste tape while you are thinking of the next word. And while there is no need to write down everything you are going to say, you should note down all the things you want to talk about, with a rough indication of how many minutes you are going to spend on each topic. Remember that the whole object of sending a tape message is to be personal, so you must learn to *talk* about your news—not make a speech about it.

Your Grundig will also take messages to—or from—other members of your own family. It is a good idea for all the grown-up members to learn how to record messages with the microphone and from the telephone.

The microphone is useful for leaving instructions—*e.g.*, what to do about meals or transport arrangements or collecting the children. This way of passing on a message has the advantage that by intonation and expression your own voice can put over information that would otherwise call for writing skill and time to express it accurately on paper.

For your messages you should always use the same spool of tape. A 3 in. message-type spool is all you need, and you should mark it with a patch of coloured marker tape. Any member of the family returning to find no one at home and the Grundig in position will know how to switch on and play back the tape. Once you get into the habit of recording messages in this way you will find that you can say what you want in much less time than it would take to write it down, and you can be quite certain that your message will be understood.

You will find the Grundig telephone adaptor (p. 154) at least as useful as the microphone for taking down messages. Leave the recorder set up permanently as for normal messages if you can, and then when a telephone call comes for an absent member of the family, someone else can always switch on the Grundig (and the appropriate input channel) and ask the caller to go ahead and record the message. If necessary the message can be played back so that the caller can check it over before hanging up. When you play back into the telephone you simply hold the transmitter end of the handset in front of the speaker. You can also use this method to pass on a recorded message to a caller when the sender cannot be there to give the message in person. And, of course, you can tie the whole thing up by getting the caller to record the reply.

Don't forget that you can use certain Grundig models with a telephone adaptor to broadcast the message you are recording to other

116

people in the room as well. You have only to turn up the monitor control while you are recording.

Musical continuity

When you have a number of recorded items on tape, you can add a little extra polish by linking them with passages of music. This gives the whole thing continuity. There are three simple ways of adding musical continuity to link up a series of recorded items on a tape: by editing, consecutive recording, or re-recording.

You can cut the tape and splice in the musical bits from another tape. Remember if you do it this way when you record the separate sounds you will have to fade them in and out, and you will have to do the same thing when you record the bits of music you are going to add. The success of this type of composite feature depends on getting it to run smoothly and keeping the gaps between each fade out and fade in short.

The next method is a little more trouble to start with, but it avoids cutting and splicing. You actually do your editing at the time of recording—*i.e.*, you fade each item in and out to a pre-determined time schedule and record the continuity music immediately after, before going on to the next sound. For most people the easiest way to add the music is from a gramophone record, but you can, of course, borrow a second tape recorder. After recording each item through the microphone you can plug the gramophone or second tape recorder into the appropriate input socket on your Grundig, start recording, and fade in the music.

Finally, if you can borrow a second recorder you can dub your recorded items on to a second tape and add the musical continuity as you go—if possible from a record player.

The great advantage of this method is that you can fade in the music as you are fading out the recorded item and avoid having a silent gap between the two. To do this you really need to use a mixer unit, although with care you can achieve much the same effect if your model has an erase cut-out button which allows you to record one item over the top of the other.

There is no need to record the music before recording the next microphone item. Provided you know roughly how much music you want to put in between each item, you can always leave that amount of tape clear—by winding on and starting the next microphone item after the appropriate interval. You can then add the music to the gaps later.

With 4-track model you can record the music on the 'spare' track, fading it in and out at the required points and then add it by playing back both tracks at once (p xx).

Noise collecting

Set a full-length tape to one side for recording any odd noises as they occur to you. You will find it invaluable for quiz games, entertaining visitors, keeping the children quiet and as a store cupboard of incidental noises for feature records, plays and creative recording—*e.g.*, making musique concrete.

117

It is a good idea to exploit the possibilities of every sound completely before passing on to the next. For example, if you decide to record the sound of your electric razor, here are a few ways of adding variations to the original theme:

Make a normal recording at 7½ i.p.s. with the razor 12 ins. from the microphone and then repeat the record at 3¾ i.p.s. Next, try a recording waving the razor about in front of the microphone and again with the razor almost touching the microphone. Now make records with the razor pressed against a tea tray, inside a tumbler, and going into action against some tough chin stubble. Remember that according to the model of your Grundig you can double or even treble the available sound effects by playing back the tape at the other speeds—*e.g.*, you can record at 7½ i.p.s. and 3¾ i.p.s. and play back each recording at 7½ i.p.s. and 3¾ i.p.s., making three different effects altogether from a single noise.

Remember, anything that makes a noise can provide interesting recordings. Here are some random suggestions for your noise collection:

Rest your watch on the microphone: it will sound like a grandfather clock. If you have a portable electric drill outfit you can create all manner of sounds—*e.g.*, run a wire brush against the edge of a sheet of metal, push different materials through the circular saw at different speeds, or hold them against the grinding wheel. Rub a wet cork against the side of a bottle; run water from the tap into various vessels; pour water from a bottle; let the air out of a balloon and then blow it up until it bursts; bend a stick until it breaks; tear a piece of cloth; rattle coins; break a pane of glass with a hammer; and so on.

Your noise collection will be lacking without a record of a thunderstorm, rain, wind (particularly the sound it makes blowing through a crack or a keyhole and through trees), waves on the beach, and running water.

To complete your collection you should work up variations on sounds like a dog whining, growling and barking, a cat miaowing and purring, bird songs.

If you have a battery operated model that lets you operate away from mains electricity supplies, you will be able to extend your noise collection still further.

'Nurse Grundig'

Don't forget that when anyone in the house is ill your Grundig can do a lot to make things easier for them by providing the right sort of entertainment as and when they want it. The invalid may have a favourite radio programme, for instance, and your Grundig can help here in two ways. First, you can use it to store the programme until the patient is well enough to listen to it (the prospect of hearing the missing instalments is in itself a valuable incentive to get better). Second, if the programme comes at an awkward time—say, just as the doctor calls—you can tape it and play it back later.

When children are quarantined for measles or similar complaints, your Grundig will bring them messages from their playmates and take back their replies.

More than once a Grundig has been used to tape either a whole church service or at least the sermon, so that it could be played back to an invalid member of the congregation. And the same idea can be made to work in reverse. Someone who is due to speak at a function or preach a sermon has an accident which prevents him from attending in person, but not from recording what he wants to say. The Grundig itself will play the tape back to quite a large audience, but for addressing crowds the output can be amplified still further through ordinary public address equipment.

Party games

There is no end to the party games you can play with your Grundig and some previously recorded tape. Apart from the ever-popular business of letting people record their own voices (at which most people's imagination dries up), there are all sorts of ways of keeping the ball rolling. Most of the ideas fall into two categories—quizzes and fun to music.

For quizzes you need previously recorded tapes, which can be of mystery noises, voices, instruments, pieces of music or quotations with some of the words erased. Preferably the noises should be preceded by an announcement; 'Here is noise number one . . .' and followed by a blank of five seconds or so. You hand out paper and pencils to the players and they have to write down their guess after each item. If necessary you can play the tape back more than once. Needless to say, you must have your own key to the noises.

This arrangement can be adapted to all sorts of subjects which you can change to suit the standards of the players. While it is useful to have at least one specially recorded tape, you can also draw on other items from your library—e.g., the family album, music collections, dance records, noise collections and so on.

Another version of the mystery voice quiz is to take your guests one at a time into the recording room and get them to read a passage in a disguised voice. You may then play the tape back and invite the competitors to identify the owners of the voices.

Musical chairs, in all its various forms, is always popular at parties and social gatherings. Your Grundig can make it even more fun since you have the pick of the radio music instead of a hard-worked pianist. For this type of game you can either record the whole thing, with surprise stops and all, or you can use a continuous record and interrupt it with the temporary stop button.

When you have a party in view you can make up a mixed dance music tape for a Paul Jones, interrupted tapes for spot competitions and similar amusements. This saves a lot of trouble on the day, and once you have started the tape you are free to join in the fun yourself.

Pets

When you have run through all the obvious uses of your Grundig in recording the voices of the various members of your family, you can turn your attention to the so-called dumb ones.

119

Your dog, for instance, is anything but dumb, and he is ready to record whines (from the other side of the door), growls and ferocious crunching noises (over a bone), joyful barks (did someone say walk?) or angry barks (postman or milkman). And for a good laugh don't forget to have your Grundig handy to record some real whole-hearted snoring if your dog is anything of a performer.

Your cat can be persuaded to contribute to your collection a varied assortment of purrs and miaows in exchange for a suitable bribe. Here again you should let the situation explain itself and talk no more than you must. Start with the microphone at a normal distance and then try to get a close-up in each case. You want the normal recording for your album, but the close-up will make excellent material for playing around with (see Noise Collecting).

Tame birds are much easier to record than birds out of doors. Parrots and budgerigars are always ready to produce tapeable sounds to order, and if you make close-up records of them and play about with the record/playback speeds you will get a collection of mystery noises that will baffle the most experienced ear. If the bird actually talks you can have even more fun. It is very difficult to get a bird to say the right thing at the right time, but with your Grundig handy you have no more worries on that score. As long as you can get the repertoire on to tape, no matter what sort of jumble it is in, you can re-arrange it (by editing) to suit yourself and any supporting script you care to write around it. The great thing is to get a really clear record at maximum volume short of distortion. To do this, especially with a soft-voiced budgerigar, you may have to place the microphone actually inside the cage; this may put the bird off for a while, so the best thing is to leave the microphone inside for a day or two before your recording session.

Caged singing birds will provide enjoyable records that can be put to a variety of uses. As all the notes are well up in the treble register, there is no need for a special microphone; the Grundig microphone will give excellent results. Good bird song records make a pleasant background accompaniment that takes the edge off awkward silences at public events like exhibitions and bazaars, and they provide a touch of novelty that attracts customers in stores, cafes and bars of all kinds. With this type of recording the output of the Grundig is ample for quite large buildings without any help from public address equipment. With about five minutes of varied song it is quite easy to produce a record of any length by simply dubbing the same section again and again on a second tape (if you can borrow a second machine).

Public speaking

Most people—and business men in particular—have to make a speech at some time in their lives, and the sad truth is that most of them do it badly. The trouble is that unless you are very, very bad your audience will put up with it; so you never get a chance to hear your faults and put them right. This is where your Grundig is invaluable.

Stand up in front of the microphone and make your speech to it. Try to imagine that you are talking to a room full of people, and once you have made a start, carry on to the end. Your Grundig will listen patiently to all you have to say and then let you hear just what it would have sounded like to an audience. But it won't flatter you; it will recall every hesitation, every mannerism and every long-winded sentence. And, fortunately, you will spot most of your faults for yourself; you won't need an expert to point them out. This is because when you *make* a speech you have in your mind a picture of how you want it to sound, and you are apt to imagine that it actually sounds that way to your audience. As a speaker you may see yourself as powerful, charming, urbane, humorous or provocative. But when, for the first time, you listen to your recorded voice, you will be able to to compare the illusion with the reality; and you will see where your actual performance falls short of your ideal. So screw up your courage and start again, trying to correct your worst faults.

If your diction is poor, try reading passages from books and plays until your Grundig gives you a favourable report. Then start on your command of words and ideas. Jot down a few sub-headings dealing with a subject you know something about and try talking for a couple of minutes at a time on each sub-heading; gradually take in more sub-headings until you can cover the whole subject without a break. You will be surprised at the difference this sort of exercise will make to your ability to talk and to your confidence in yourself. The improvement is something that will affect your ordinary conversation and your personality as well as your skill in proposing a motion.

All this doesn't mean that your Grundig can take the place of a qualified teacher. If you are going in for public speaking seriously it will pay you to learn the right way; take a course on the subject and join a public speaking group or a debating society. You will probably find your teacher uses a Grundig himself and you will certainly find your own useful for the homework he sets you.

Radio

With an FM radio receiver or tuning unit connected to your Grundig you open a door into an entirely new world of possibilities. For more and more people this is the real reason for owning a Grundig. You simply leave it permanently connected to the radio and loaded with a tape so that it only needs to be started when an item turns up that you would like to preserve. With some Grundig models you can also use the tape recorder amplifier itself to amplify and reproduce the signal from a radio tuner for all your normal listening. With this arrangement you hear the radio through your Grundig and simply press the right control when you want to tape it.

So long as you only want to record radio programmes to entertain your own family and friends you aren't likely to run into copyright trouble, but beware of using your recordings for any other purpose— *e.g.*, at a local church bazaar or a club dance.

Provided that you have a good aerial and tuner, the quality of your recordings from the FM radio broadcasts will be in every way equal to the original broadcast. Moreover, you can afford to record everything you listen to on the chance that you might want to hear all or part of it again. And if something you want to hear comes on at an awkward time, you can take it down on tape and play it back at your leisure. If you don't want to listen as you record the programme you can turn down or switch off the speaker(s); this is useful, for instance, if you want to tape a programme after the children have been put to bed. And with most Grundig models, as long as you can switch on at the beginning of the programme—or get someone to do it for you—there is no need for you to be there to switch off; the automatic stop will switch off the motor when it comes to the end of the tape.

FAMOUS VOICES. Some tape enthusiasts set aside a tape for recording a collection of the voices of famous people—the Royal Family, Cabinet ministers, actors, musicians, conductors, film stars, sportsmen, explorers and the like.

It is always best to cut—or dub—these examples from a longer recording so that your tape will present characteristic glimpses of each personality without taking a long time to do it. A few seconds of each voice are all you need to make an interesting tape. Note the tape positions of each item in your tape index and leave two or three seconds of blank tape between each item. Don't record anything in your own voice to identify the speaker and you will be able to use this tape also for your mystery voice contests.

MUSIC. If you are fond of music and want to build up a library of the works that interest you, the Grundig-radio set-up is all you need. Look up the radio programmes in advance and decide what you want to tape. Check the programme times to see exactly how much tape you are going to need, but don't cut it too fine; remember to allow spare tape to record the sound of the orchestra tuning up, the preliminary announcement (which often includes useful programme notes) and the applause at the end. These are parts of the programme that you do not get on disc recordings, but they add an atmosphere of reality to the performance that heightens the pleasure of listening. When you tape important works in this way it is a good idea to cut out the reference from your printed radio programme and stick it in your tape index along with any press criticisms of the performance.

Even if you don't intend to keep a permanent tape of a particular work, there is a lot to be said for recording it temporarily. When you have it on tape you can listen to it and study it at leisure. If you are interrupted by the telephone or a visitor, you can stop the tape and carry on listening later, whereas if you are listening to the actual performance, interruptions are fatal. You can study unfamiliar works and composers in this way and greatly enlarge your musical appreciation. Another advantage is that you can compare various interpretations of the same work and keep checking your library copy of a favourite work against new performances by different players and conductors. You are then always in a position to scrap your library copy and replace it by a performance that you prefer after careful comparison.

When you want to tape a work that extends beyond one complete tape track, you have to plan your recording session so that you change over to the second track between two movements. To make a quick change-over, it is best to have a second pair of spools already threaded. When you want to change, stop the tape, lift off both spools, drop the new spools into position and start the tape again.

If you want to make tapes of the highest possible quality for playing through external high fidelity equipment (p. 48), you should record at 7½ i.p.s. However, for recording music for dancing and party games —*i.e.*, anything that you will not be listening to critically—a speed of 3¾ i.p.s. is all you will ever need, and it makes your tapes go twice as far.

PLAYS. You can use your Grundig for taping radio plays so that you can listen to them at a time to suit yourself. Many busy people prefer to listen to plays over mealtimes. When you serve up plays in this way you can always have an interval to suit your domestic arrangements—a thing that is impossible when listening to the original broadcast.

Taped plays are popular items at women's sewing groups, and they offer a perfect solution to the problem of letting youngsters hear a programme that is transmitted after bedtime.

RECIPES. Some daily radio features include a cooking recipe. Even if this is read at dictation speed it is always easier to tape it and play it back when it is wanted in the kitchen later. If it turns out to be worth preserving, the details can be entered in the recipe book; otherwise it is simply left to be wiped out by the next recording.

VARIETY. There is usually one item in a weekly variety feature that you like better than the rest of the programme. Your Grundig lets you pick out these plums and make a collection that you can play back for fellow-enthusiasts to enjoy.

Have you a favourite comedian? You can make up a tape with excerpts from every performance he appears in and then edit it to give you the best of the bunch.

You can also record snatches of singers, comedians, actors and compères for mystery voice competitions.

Speech training

Because your Grundig can take down not only what you say, but how you say it, you can use it for studying and improving your diction, your voice and your manner of talking. All these things are an important part of your personality; they influence other people's opinions of you and how much attention they pay to what you say.

Like most people, you probably take it for granted that you have a pleasing voice until you hear a tape recording of it. If you are prepared to admit your faults, your Grundig will soon help you to do something about it. Bearing in mind the voice you thought you had, play the recording back again and note the things you dislike about it; monotonous pitch, slovenly pronunciation, fading out at the end of a sentence, etc. Now make a second recording and try to do better (leave the previous recording there for comparison). You will be surprised what a big improvement you can bring about in a short time.

One word of caution. If you have a ribbon microphone and talk close to it, say within 6 ins., it will tend to flatter you voice, making it sound rich and deep-toned. This peculiarity is much exploited by crooners and radio actors. There is no harm in using the trick when you are recording so long as you realise that your own voice doesn't really sound as well as that. If you are concerned with improving your normal voice, you should keep the microphone between 1-2 ft. away; at this distance the ribbon microphone will give faithful reproduction.

123

Staff selection

In business your Grundig can help you to pick the right man for the job. When you have to interview a number of applicants it is often difficult to remember anything of the earlier interviews, and the more you try to mend matters by taking notes, the less time you have for studying the subject of them.

If you tape each interview on your Grundig you can go ahead and ask your questions while you study the applicant. You need only take a few notes for reference because you can check back to the actual interview whenever you like. It does not matter whether the applicants know they are being recorded. Some employers prefer to conceal the equipment because it might make the applicant feel nervous; others deliberately leave it open for the same reason.

Tape friends

If you want to exchange tape messages with other enthusiasts in other towns—or other countries—you can join one of the many tape correspondence clubs. These are organised to introduce people to each other through the medium of tape, so that they can exchange ideas and tape recordings. In some clubs, tapes are circulated to members who add their comments to the tape and pass it on to the next on the list. Members can submit tape features for postal criticism by the rest of the club. Other clubs put individual correspondents in touch on the lines of the various 'pen pal' organisations.

Either through organised club activities or through individual contacts made in this way, you can use your Grundig to add interest to your hobbies—*e.g.*, if you are a keen bird watcher, you can exchange bird-song records with others and swap yarns about your natural history excursions; if you are an amateur musician, it gives you a chance to compare your own efforts with another musical friend on the other side of the country—or the world. You can even play a duet with your correspondent. It is easy: each of you records his own part of the arrangement. You then exchange tapes and play back the recording while you add the other part on your own instrument (if you have a friend with a second recorder you can also tape the result of your joint effort). With a 4-track model (p. 35) it is of course easier still, because you can make each record on a separate track on the same tape and then play them back together.

Your tape correspondence doesn't have to stick to one language. If you are studying a foreign language you can carry on a very useful tape exchange with a fellow-student of your own language in the other country. In this way you can get—and give—expert criticism and profit by first class examples that you can play back again and again until your own pronunciation is perfect.

Teaching

Nowadays a tape recorder is considered essential for practically every branch of education and instruction. The teacher with a Grundig can make lessons more interesting for the class and easier for himself.

DANCING. Apart from providing music for dancing and dance instruction, a Grundig is excellent for use in teaching tap dancing. It allows pupils to listen to their mistakes as often as the teacher cares to play them over. And it is ready to bring an orchestra to recitals when the studio piano would sound thin.

DRAMA. There are all sorts of ways of using a Grundig in voice production and elocution. The standard method is to record a 'model' passage spoken by a professional and then to get the pupil to record the same passage immediately after. The pupil can then spot the faults in his own performance and he can work with the teacher to correct them. By leaving each successive effort on the tape the pupil can check progress and gain confidence. Without actual proof on tape he would have to rely on his memory.

Group practice in reading plays also can be taped and played back for criticism, giving the students valuable experience in arrangement and production. It can also help actors to learn their parts word for word.

GENERAL EDUCATION. Almost every school possesses at least one tape recorder, and articles on new uses for them in the classroom appear regularly in the educational press. A detailed list of the ways you can use a Grundig for teaching would easily fill a book, but the same general principles apply to all the uses—i.e., providing the class with examples to copy, recording their efforts for criticism and comparison with the models, and taping the best examples of individual and group work for the school tape library.

In addition, a Grundig is useful for taping broadcasts to schools so that they can be played back to classes unable to hear them at the time. Some teachers have gone so far as to tape lessons and talks to send to pupils in hospital so that they will keep in step with the rest of the class.

LANGUAGES. With your Grundig you can considerably cut the time it takes to learn (or teach) foreign languages. One of the most useful ways of acquiring a vocabulary is to select a foreign language item, such as a news bulletin or talk, broadcast just too fast for you to understand. Play the tape back phrase by phrase until you have the meaning and then listen to it straight through once or twice until it is familiar. Keep doing this and presently you will find that you have increased both your speed and your vocabulary.

If you are teaching a class you can get each pupil in turn to concentrate on a phrase and translate it so that the whole class gets the benefit. When you come to pronunciation you can use the foreign broadcasts again as object lessons to play over and over until your pupils are perfect—as checked by recording their best efforts. For beginners you can tape the news bulletins given out at dictation speed by foreign stations on short waves.

MUSIC. Whether your pupils are learning to sing, or play instruments, you can get on much faster with your Grundig to provide up-to-date examples of the work of leading performers for study.

You can use it in conjunction with your radio to collect illustrations of different styles of playing or types of interpretation; you can tape the pupil's own efforts and compare them with those of an experienced performer, or play them back for the criticism of the class; you can make permanent recordings of school concerts that would otherwise be forgotten. Pupils are much keener when they can actually listen to their own music, whether it is an individual effort or the work of a group. When you come to think of it, the only way a member of a string quartet can really know what the general effect sounds like is to hear a record of it played back.

CHAPTER 8

How To Do It Better

YOUR GRUNDIG IS REALLY TWO INSTRUments in one. It is first and foremost a completely self-contained recording and reproducing machine that you can take around with you wherever you go. But you can make it the starting point for a more ambitious high fidelity set-up by adding other pieces of equipment.

You are probably familiar with the term hi fi, even if you are vague about its meaning. It indicates (or should indicate) a standard of reproduced sound which is a very close imitation of the original—not just in somebody's opinion, but in a way that can be actually measured and compared.

Frequency response

All sounds are caused by vibrations in the air—vibrations that radiate away from the sound source in pressure waves, like ripples on the surface of a pond when you drop a pebble into the middle. The slowest vibrations are the low notes and the quickest are the high ones. Each vibration goes through a complete cycle starting in the middle of the wave, rising to the crest, falling to the trough and coming back to the middle again. You can measure how many of these complete cycles occur every second in any particular note of music; the number per second is called the frequency of the note. There are 16 cycles every second in the lowest note of the organ, and 16,000 per second in the highest note given out by a violin. Many of the instruments in the orchestra create vibrations that contain even higher frequencies—up to 25,000 per second and more.

For high fidelity reproduction you aim at reproducing all the musical frequencies in the same proportion—*e.g.*, if you record a note at 1,000 cycles and the sound when you play it back is exactly half as loud as the original note, then any other frequencies must be reproduced at exactly one half the original strength. (You measure loudness of sound in decibels. One decibel approximates to the smallest step in the loudness scale that the trained human ear can detect.)

If you draw a graph of this type of reproduction, and the loudness (measured up and down) stays the same as you go from the lowest to the highest frequencies (measured

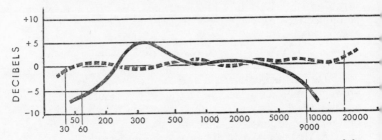

FREQUENCY RESPONSE. *This type of graph gives a picture of the way your Grundig reproduces the range of sound frequencies* (horizontal scale) *in terms of acoustic power* (vertical scale). *Typical tape recorder response* (full line) *is within ±5 dB from 60 to 9,000 cycles per second. Typical amplifier response* (interrupted line), *±1 dB from 30 to 20,000 c/s.*

from left to right), you get a straight line running horizontally from one side of the picture to the other. If any frequencies are reproduced louder than the others the curve will rise in a hump at these particular frequencies. This type of picture of the way the various frequencies are reproduced is called the response curve of the equipment. Manufacturers of sound equipment always state its performance in terms of its response, from so many cycles at the low end of the scale to so many at the top end. And if the response is said to be flat from, say, 50 cycles to 12,000 cycles, you know that it will reproduce all frequencies faithfully between those two limits. And you would know that a piece of similar equipment with a flat response from 20 cycles to 20,000 cycles should be of an even higher standard and would probably cost much more.

Now if you are designing a tape recorder, reproducing high frequencies doesn't give you much trouble; but when you want to handle the really deep notes, like the tuba, double bass and the big drums, you are up against the problem of size. You can't do justice to the low frequencies unless you have a big loudspeaker capable of passing on its vibrations to a large volume of air. This is why the instruments just mentioned are not only the deepest sounding in the orchestra, they are also the biggest.

So if you want your tape recorder to be small enough for you to carry around, you have to sacrifice some high fidelity at the low end of the scale. Most Grundig tape recorders are fitted with elliptical speakers which give better low frequency reproduction than the largest circular speaker that would go into the same carrying case, and some models have more than one speaker. For real concert quality reproduction there is no substitute for a bigger speaker unit—but of course you can't carry it about with you.

However, your Grundig will *record* all the low notes even if it can't reproduce them all to the best advantage. So when you connect it up to a bigger speaker unit, you get the full benefit of the extra frequencies recorded on the tape. These suddenly make themselves heard and the reproduction of orchestral music and many other sounds appears so much fuller and richer.

Most Grundig dealers can demonstrate a variety of speakers that you can use in conjunction with your model in this way. The choice is extremely wide and the only reliable test is to try them all out and if possible make your final selection at home under your normal listening conditions.

The speaker must be mounted in a suitable cabinet or on a large heavy board called a baffle. This is to prevent the opposing sound waves sent out by the front and the back of the speaker from simply cancelling out. The baffle board does this, but the cabinet goes further; it collects the sound waves from the back of the speaker, turns them around and shoots them out of another opening in the cabinet so that they add themselves to the waves at the front. For this reason it is called a reflex cabinet.

You can make a suitable cabinet very economically by standing a heavy board across the corner of the room and fitting a triangular top to it. You can also buy corner units

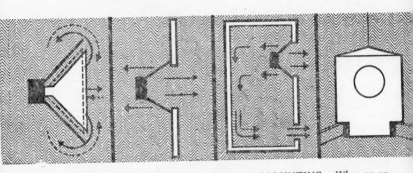

TYPES OF MOVING COIL SPEAKER MOUNTING. When an un-mounted speaker cone (left) *vibrates at low frequencies, the sound waves from the back and front cancel out* (shown by arrows). *A rigid baffle gets over this but must be large to be effective. A reflex cabinet adds the sound waves from the back of the speaker to those from the front. A corner housing* (right) *is the easiest way to fit a reflex cabinet into a small living room.*

of this kind ready-made with the hole for the speaker and the reflex ports already cut. But if you construct your own cabinet you must use heavy materials that will not vibrate at the frequencies you want to reproduce. If you can't get thick enough wood, frame two thicknesses of light plywood to make a gap (about 1 in.) between them, and fill this with sand or cement.

Your extension speaker must be of the right impedance to match the output of your tape recorder and must be fitted with a suitable lead and plug. (You need a 3 ohm speaker for all models except the TK830/3D which has a 15 ohm output for the extension speaker.) Remember that you want all the output of your tape recorder to go into the extension speaker, so you will have to switch off the built in speaker. On some Grundig models this is taken care of automatically as soon as you plug in the extension speaker; on others you can cut out the speaker independently.

Since the extension speaker brings out low frequencies that you couldn't hear through the speakers built into your tape recorder, it will appear to increase the volume; but in fact, it is using exactly the same power as before.

What's watts?

Watts, as you know from your ordinary domestic equipment, measure power. The power of a lamp or a radiator is measured in watts, and the power of your Grundig output is also given in watts. Depending on the model, it may have an output of anything from $2\frac{1}{2}$ to 6 watts 'undistorted' —that means without any noticeable distortion of the natural sounds from which the record was made. This volume of good quality sound is about as much as you can expect from portable equipment where size and weight put definite limits on the performance. And in the average room an output of $2\frac{1}{2}$ watts gives you a very satisfactory volume of reproduction.

But you get a different answer if you look at the problem from the other end—*i.e.*, how much sound do you want in order to produce the same impression that you would get from listening to the actual orchestra, singer or speaker. This is something that you can measure, and although the power varies according to the source of the sound and the surroundings, you will find that to fill the average living room you want at least 10 watts to play with—some people say two or three times that amount. So for true 'orchestra-in-the-room' reproduction, it isn't enough to add an extension speaker; you have to increase the actual power that goes into the speaker. You can do this with an external amplifier.

The external amplifier

An external amplifier is similar in principle to the amplifier built into your Grundig. But the built-in amplifier is

131

designed to amplify a very weak signal coming from the sound head so that it will give a reasonable volume of sound from the loudspeaker. To do this it has to multiply the signal very many times. The external amplifier usually has a much more powerful signal to start with.

You buy an external amplifier as a complete unit; you connect it to the mains, then plug into your Grundig high impedance output socket and connect the amplifier output to a loudspeaker unit.

You have a very wide choice of amplifiers and the answer is the same as with other equipment: try all the likely models within your price bracket, and if you can, make your final decision by testing the complete set-up—Grundig, external amplifier and loudspeaker unit—in your own home. Up to a point, the more you pay, the better the outfit you will get; but there is no sense in paying for a 55 watt amplifier if you live in a flat with thin walls. And remember, too, that no matter how much you pay for quality in the amplifier, you can't add anything to the frequencies recorded on the tape itself. So if your Grundig is one of the simpler models with a single speed of $3\frac{3}{4}$ i.p.s. and a flat frequency response from, say, 60 to 9,000 cycles per second, you would be wasting your money if you bought an amplifier with a very much better specification. Remember, too, that the output power of the amplifier should always be equal to or less than the maximum undistorted power that your speaker unit can handle.

The pre-amplifier

You can connect the output of your Grundig straight to the input of the external amplifier and it will work quite well—provided that the impedances match (p. 139) and also that the strongest signal coming from the tape recorder does not overload the input stage of the amplifier. But as a rule, when you have gone to the expense of an external amplifier and a separate speaker unit, you will want to use them for reproducing from other sources too—e.g., a gramophone pick-up, a radio tuning unit or a microphone. All these

132

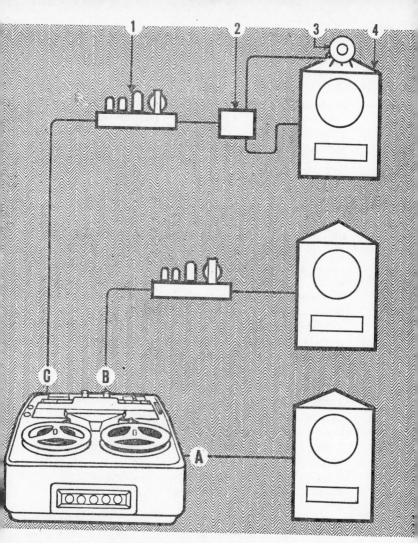

BETTER WAYS. You can improve low frequency reproduction by using an extension speaker in a reflex cabinet (A) and increase the power with an external amplifier (B). A high fidelity set-up (C) has external amplifier (1) and cross-over filter (2) which splits up the signal between high frequency (3) and low/medium frequency speakers (4).

133

sources have different characteristics—some give a strong signal and others a much weaker one, some may have a high impedance and some a low.

All these pieces of equipment talk a different 'electrical language' and your amplifier can only understand one. So you have to use a kind of universal interpreter that can translate all the different types of signal into the only one your amplifier can understand. This is a unit called a pre-amplifier; it is connected between the amplifier and all the signals you want to reproduce.

The pre-amplifier normally has various inputs for you to plug in a gramophone pick-up (more often two, for 33⅓ r.p.m. and 78 r.p.m. records), a radio tuning unit, a microphone (usually both high and low impedance types) and a tape recorder output (so that you can play back the tape through the amplifier). There will also be controls for adjusting the volume of the signals and boosting them if necessary, and other controls for adjusting the amount of bass and treble.

So you see there is quite a lot of work for the pre-amplifier to do. But once you have everything connected up to it and you plug into the external amplifier, you have complete control over all the signals without doing any further fiddling. By turning a knob or a selector switch on the pre-amplifier you can choose any signal source to record from or play directly through your sound equipment, and you can control the volume and quality to suit your taste.

But don't think that all this extra equipment is necessary before you can enjoy yourself with your Grundig. Without any of these extras you can still make excellent recordings direct from your radio, gramophone, microphone and other sources, and the amplifier and speaker of your Grundig will reproduce them with all the quality and volume you need for normal domestic listening. The extra equipment is for the connoisseur who is prepared to go to a great deal of trouble and expense to get the ultimate in high fidelity. And don't forget that one of the great attractions of a portable tape recorder is that it *is* portable, and gives you

134

a lot of pleasure that you would miss if you couldn't take it around with you.

More speakers

If you want to get down to the really low notes, you will want anything from a 10 to a 15 in. speaker in a reflex cabinet with up to 9 cubic ft. of enclosed air space. This class of speaker unit will do full justice to the low notes of the organ, the double bass fiddle and all the other deep voices of the orchestra. But you can't expect the same loud-speaker to perform as well on the high notes—any more than the same singer can tackle bass and tenor parts. So you have to add another, smaller speaker to handle the upper register—and if you are really critical you may appreciate a third unit (called a 'tweeter') for the extremely high frequencies given out by such instruments as the triangle, cymbals and other special effects.

Fortunately, these extra speakers don't need the generous accommodation called for by the bass speaker (usually referred to affectionately as the woofer). Frequencies above about 1,000 cycles can be handled quite well by the speaker cone alone, and the bare minimum of a surround will serve.

When using an arrangement of two or more speakers in this way, you should not connect them straight to the amplifier or they won't share the work in the right proportion. Instead, you have to connect the speakers to the set through a special distributing device called a crossover filter. This splits up the signal into as many frequency bands as you have speakers to feed, so that each speaker handles only that range of frequencies it can deal with best.

A typical set-up would be three speakers: a 12 or 15 in. speaker in a reflex cabinet, with an 8 or 10 in. speaker in a simple frame and a 3 in. speaker mounted on top. The crossover filter would feed all frequencies below 1,000 c.p.s. into the first speaker, everything from 1,000 to 3,000 c.p.s. into the second, and everything over 3,000 into the third.

As you go higher up the frequency scale, the sound waves from the speaker cone tend to become more and more directional. If you stand right in front of the tweeter and then move a few feet to either side, you will notice a rapid falling off in the upper frequencies. This effect is known as beaming. There are various ways of getting around it, such as directing the sound through short horns arranged in a fan shape, pointing the speaker into the corner of the room to get reflection from both walls, or pointing the speaker up into a reflector, shaped to direct the sound out into the room.

If you own the Grundig model TK830/3D you can add a Grundig distributor speaker unit (p. 156) which consists of two high frequency speakers mounted in such a way that the sound is non-directional. This unit lends crispness and brilliance to the reproduction and enhances the three dimensional effect given by the internal elliptical speakers. The distributor speaker plugs into a special socket in the distribution panel.

Loudspeaker types

There are several types of speaker for you to choose from, and it will help if you know their principal characteristics before you start.

MOVING COIL SPEAKERS. These are by far the commonest types in use today. They are fitted in all Grundig models. The driving unit is a coil of wire (known as the speech coil) wound on an extremely light cylindrical former and supported in the field of a powerful permanent magnet. The magnet is shaped so that all its lines of force are concentrated in a narrow circular gap in which the coil is suspended. The coil is free to move parallel to the gap but not sideways, and is fastened to the apex of a cone of stiff fabric mounted in a circular frame. (The speakers fitted to Grundig tape recorders are made elliptical with the object of getting the biggest possible cone into the case without making it cumbersome.)

136

The amplifier output is connected to the speaker coil. As the signal flows through the cone it creates an alternating magnetic field. This makes the coil vibrate to and fro in the

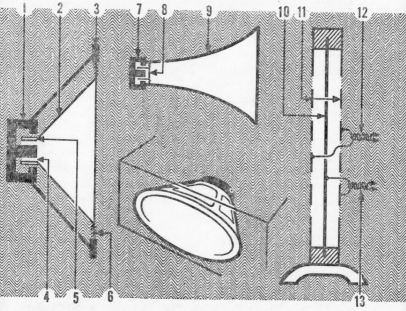

SPEAKER TYPES. Moving Coil (left): *1 Permanent magnet. 2 Cone. 3 Chassis. 4 'Spider' holding coil central. 5 Coil. 6 Flexible suspension.* Horn (top centre): *7 Permanent magnet. 8 Coil. 9 Horn.* Electrostatic (right): *10 Diaphragm. 11 Polarised screens. 12 Signal leads. 13 D.C. supply.* Elliptical (lower centre): *form of moving coil speaker used in most Grundig models.*

field of the permanent magnet and its movement is transferred to the cone, making it send out sound waves corresponding to the original signal.

The sound waves go out equally from the front and back of the cone and one of the problems of mounting a moving coil speaker is to transform all the energy fed into the coil into useful sound—which means harnessing the vibrations from the back as well as the front of the speaker cone. This is the main purpose of the cabinet or baffle mounting.

Moving coil speakers have been perfected over so many years that in practice their performance leaves little to be desired.

HORN LOADED SPEAKERS. A horn loaded speaker is just a special adaptation of the moving coil type, above. Theoretically a horn is an excellent device for transforming the energy in the speech coil into useful sound. But in practice there are two snags: the shape of the horn must increase in cross section according to a complicated formula, which makes its construction expensive; and it needs to be at least 20 ft. long to reproduce the low frequencies satisfactorily. The size can be reduced by folding the horn in various ways, but even so, this type of speaker is not for the ordinary living room.

Some manufacturers make a special type of multiple horn speaker for handling the high frequencies. In this case the horns are only a few inches long and they are shaped so that they spread the high frequencies instead of beaming them over a narrow angle—an undesirable trick of the ordinary moving coil speaker.

The horn type of speaker is used principally in cinema sound installations and on public address systems.

ELECTROSTATIC SPEAKERS. In theory this type of speaker gets closer to the ideal than any other. The actual driving unit is a thin metal diaphragm, up to several feet square, which vibrates as a whole in response to the signal. The diaphragm is mounted in a frame between two perforated metal plates. The signal is fed to the outer plates and a polarising voltage is applied to the diaphragm so that an electrostatic force is set up between the diaphragm and the outer plates. The fluctuations of the signal cause the whole diaphragm to vibrate to and fro and create sound waves.

This type of loudspeaker gives a particularly pure and undistorted reproduction of the original sound—almost like listening to the real thing through an open window the same size as the speaker. An electrostatic speaker simply stands on the floor a couple of feet away from the wall; it does not need any type of acoustic enclosure or special mounting.

Electrostatic speakers usually cost more than moving coil speakers of comparable performance and they require extra components to provide the polarising voltage for the diaphragm.

Small electrostatic speakers are manufactured to work as tweeters in a normal two or three speaker assembly. Electrostatic tweeters cost about the same as moving coil speakers of the same class. Which you choose will depend on your particular taste. For that reason you should always ask to try out the unit in your own home with your own equipment.

Connecting equipment

Before you connect anything to your Grundig (or any other make of tape recorder for that matter) you must be sure that its electrical characteristics are suitable. The two most important electrical characteristics are impedance and signal level.

Unless you are using Grundig accessories, in all probability you will also have to connect the right type of plug to suit the Grundig sockets.

THE IMPEDANCE. In an electrical circuit, impedance measures the opposition a circuit offers to a change in the current flowing through it; in particular, to an alternating current, since that is changing all the time. Impedance is measured in ohms, just like resistance.

Electronic accessories for sound reproduction—e.g., microphones and speakers—are divided roughly into high and low impedance equipment. An impedance of less than 600 ohms is regarded as low and anything over about 2,000 is high. The input and output sockets on your Grundig are classified as low or high impedance. All Grundig models have a low impedance output and most have high impedance outputs as well. Nearly all models have high impedance microphone inputs, although some Grundigs have low. Your Grundig dealer or the data sheets at the back of the book will keep you right here.

The general rule is that you should always connect low to low and high to high impedances. On the input end you need be no more accurate than that, since there is no actual transfer of power from one side to the other.

When you are dealing with power outputs—*e.g.*, when you want to connect an extension speaker to your Grundig output—you have to match the impedances more exactly. So if you want to connect an extension speaker to, say, Grundig model TK8, which has a low impedance output of 3 ohms, you should use a low impedance (3-15 ohm) speaker.

THE SIGNAL LEVEL. You also have to worry about signal levels—(1) the level of the strongest signal that will be available from, say, a microphone or a pick-up, or at the output of your Grundig and (2) the level of the weakest signal that can be handled by the input socket of a tape recorder, amplifier or similar equipment. This level is given in the specification in volts or millivolts (mV).

In an amplifier the level of the output given is the maximum voltage developed when the unit is fed with the strongest signal that it can handle without distortion. With a microphone the figure given is usually the voltage developed by a pressure of 1 dyne per sq. cm., which is approximately what you produce when speaking normally into the microphone from a distance of 3 ft.

Here the rule is that the signal level at the output of the equipment you are connecting must not be greater than the signal level required to load the input of the next piece of equipment fully. This is simply a matter of adjustment of the output volume control. Of course, the signal level must not be so low that the amplifier cannot step it up to a reasonable volume.

INPUT AND OUTPUT CONNEXIONS. All input and output connexions on the Grundig models are made through screened cable terminating in either 6 mm. jack plugs (on the early models) or DIN-type plugs with three or more pins, depending on the model. (For information on which models use which type of plug, see the data section at the end of the book).

The 6 mm. jack plug consists of a hollow metal sleeve, 6 mm. in diameter, with an insulated conductor running

140

down the centre, terminating in a metal tip. Inside the plug body, the tip is soldered to the signal wire and the metal braid covering on the screened cable is connected to the sleeve of the plug. When the plug is pushed into the socket

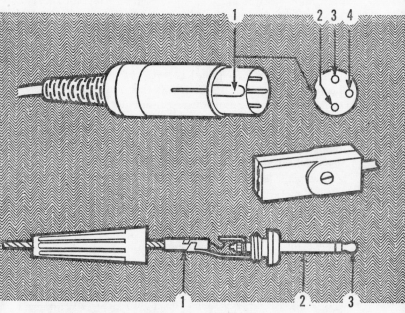

PLUG TYPES. Bottom: Jack Plug *used on earlier Grundig model inputs and outputs. 1 Sleeve tag and cable grip connected to screen. 2 Sleeve 3 Tip (connected to signal wire).* Centre: *One type of* Mains Lead Connector *for connecting mains lead to Grundig on models where lead is detachable.* Top: Three-pin Plug *used on later Grundig models inputs and outputs.* Top right: *Connexions 2, 3 Signal wires. 4 Common, screen or earth. 1 Keyway to locate plug.*

on the distribution panel the tip and sleeve make contact with the input circuit through spring loaded contacts in the socket. The tip of the plug may also operate separate sets of contacts in the socket and thus make any necessary changes in the associated circuits. On all plugs the sleeve is always earthed and the tip is connected to the insulated wire. All jack plugs have a recess below the tip which en-

141

gages a spring-loaded finger in the jack socket and holds the plug in position.

The British standard jack plug has a sleeve diameter of ·25 in. This is just too large to go into a 6 mm. socket, but on Grundig mixing units which have either jack plug sockets only (GMU 2) or duplicated jack plug and three-pin sockets (GMU 3) the sockets have been made big enough to take the ·25 in. plug as well as the normal Grundig 6 mm. type. The GMU 1 mixer unit has 6 mm. sockets.

The three-pin DIN plug has three contact pins spaced at twelve, three and six o'clock and surrounded by a metal sleeve. When the plug is pushed into place a keyway in the metal sleeve has to register with a key in the socket, so that you always fit the plug the right way round. The pins extend through the insulated mount of the plug to solder tags at the back where they are joined to the connecting lead. The centre pin is always connected to the screen of the lead and to the metal sleeve of the plug. The other two pins are connected to the inner insulated wire or wires.

Connecting your Grundig to the mains

Whatever else you may want to connect to your Grundig, you will have to connect it to the main electricity supply. You do this with the twin- or three-core flexible lead supplied with your machine. The lead may be already attached to your machine and stowed in a pocket next to the distribution panel. Or it may be detachable and stowed either under a flap on the front of the deck or in a compartment sometimes situated in the lid. When the lead is stowed separately it has a plug at one end which fits into a socket on the side of the recorder or on the distribution panel near the fuse holders.

Two-Core Leads. If your Grundig is one of the models with a two-core mains connecting lead it should be fitted with a two-contact plug. This may be a b.c. (bayonet cap) adaptor designed to plug into an ordinary domestic lamp holder, or it may be a two-pin plug to fit into a socket

142

outlet on the wall or skirting board. If you fit a b.c. adaptor and you want to plug it into a two-pin socket you can buy an adaptor which has a socket to take your b.c. fitting and two pins to plug in to the supply socket. It is a good idea to have one of these adaptors anyway, because it may not always be convenient to use a lamp socket. You can also get a two-way b.c. adaptor that you can plug into a lamp holder to take both your Grundig lead and a lamp.

If your electric points are all of the three-pin type and it is not convenient to use a lighting socket, there are two ways of connecting up your Grundig: (1) there will almost certainly be an odd reading lamp available that is already wired to a three-pin plug: you need only plug this into your three-pin outlet, take out the light bulb and plug in your Grundig, or (2) you can make up an adaptor with a length of twin lighting flex, a b.c. lamp holder and a three-pin plug; you connect one end of the two flexible wires to the two terminals of the lamp holder and the other to the terminals of the three-pin plug marked L and N.

THE EARTH CONNEXION. There is a socket for an earth plug on the distribution panel of most Grundig models. These models are perfectly satisfactory in normal use without an earth connexion, but for the highest quality reproduction—e.g., when you are taping an orchestral programme or when you are playing back through high fidelity equipment—you should always connect up the earth socket to a reliable earth point. For this you can use a Grundig jack plug Type J6 with a suitable length of heavy flexible wire and connect the earth socket of your Grundig to the earth socket of your electricity supply (through the earth pin of a three-pin plug). If you are not using a three-pin plug, you can employ the earth connexion of a radio or T.V. set; a cold water main pipe also makes an excellent earth point, *but do not connect to a gas pipe*.

If there is no special socket provided for an earth connexion you can take it that your model is earthed through the mains lead or that it doesn't need an earth connexion at all.

THREE-CORE LEADS. If your Grundig has a three-core mains lead you should, if possible, connect it to one of the standard types of three-pin plug. This will be either a 2 or 5 amp, three-pin plug (for normal domestic three-wire systems) or a 13 amp fused plug (for ring mains).

To wire up any of the above three-pin plugs: connect the green wire to the earth terminal, connect the red wire to the left-hand terminal (looking at the plug with the pins away from you and the earth pin underneath) and connect the black wire to the right-hand terminal. When you wire the plug in this way your Grundig will be automatically earthed as soon as you connect it to the mains.

If your domestic points are of the two-pin type, or if you particularly want to connect your lead to a lamp socket,

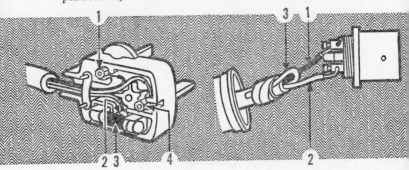

HOW TO CONNECT THE THREE-CORE MAINS LEAD. (Left) *When you fit a three-core lead to a three-pin plug, be sure to connect the green wire to the earth terminal (4), the red wire to the live terminal (2) and the black wire to the neutral terminal (1). Check fuse (3). (Right To fit a (b.c. adaptor to a three-core lead, turn back the green wire (3) and connect red and black wires (1 and 2) to the plug terminals. There is no right or wrong way, but you may have to change the plug around in the socket if you hear a hum.*

you can ignore the earth wire and fit a b.c. adaptor or two-pin plug as described above. In this event, the green earth wire should be cut short so that it does not go into the plug; better still, bend it back on itself and bind it to the flex with insulating tape.

CHAPTER 9

Extras to help you

YOUR GRUNDIG IS A COMPLETE UNIT IN itself, and without any further assistance it will do everything you expect of a good tape recorder. But if you are a real tape enthusiast you may want to extend its scope still further; to help you there is a range of Grundig accessories and extras specially designed to work with your machine. These accessories are described below; you can buy any of them from your regular Grundig dealer.

Always check that any accessories you buy to connect your Grundig are the right ones for your particular model: they are not all interchangeable (see Connecting Equipment, p. 139).

The Grundig microphones

The microphone is the ear of your Grundig. It listens to the sounds you want to record and changes them into electrical vibrations which it sends down the flexible lead into the tape recorder amplifier. This is more or less what your ear does when it converts sounds into impulses and sends them along the auditory nerve to your brain.

Most Grundig models are sold complete with a general purpose microphone; some are sold less microphone and you are free to choose the one that suits your purpose.

The general purpose microphone may not give the best results with special subjects that interest you. So you want to know something about what kinds of microphone there are and which subjects they record best.

To make a choice of microphone you need to know something about the various types available. Remember, there is no point in buying a particular microphone simply because it costs more; it may not give you noticeably better results with your model. The best microphone for you is the one that does what *you* want with *your* equipment.

There are four types of Grundig microphone: moving coil (dynamic), ribbon, condenser and crystal.

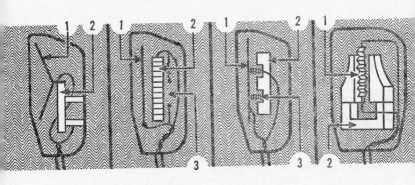

THE MICROPHONES (reading from left to right)—Crystal: *Diaphragm (1) picks up sound vibrations and conveys them to a crystal element (2) which converts the movement into an electrical signal.* Condenser *microphone: Sound waves strike diaphragm (1) and move it in relation to stationary plate (2) charged to 100 v. through lead (3). Changes in electrical capacity provide the signal.* Moving Coil (*Dynamic*): *Diaphragm (1) causes coil (3) to vibrate in field of magnet (2) generating electric current in coil to form signal.* Ribbon: *Sound waves make metal ribbon (1) vibrate between poles of powerful magnet (2) generating voltage difference between ends of ribbon and causing current to flow in microphone leads.*

MOVING COIL (DYNAMIC) MICROPHONES. The moving coil type of microphone belongs to the pressure type— *i.e.*, it has a diaphragm which is moved by the pressure of the sound waves acting on one side of it. The diaphragm is connected to a coil suspended in the field of a powerful permanent magnet, the arrangement being similar to that used in a loudspeaker (p. 136). When the diaphragm moves in response to sound vibrations, it pushes the coil to and fro in the magnetic field and sets up electrical vibrations which form the 'signal' you record on your Grundig.

Dynamic microphones are excellent for speech and singing and for all but the very highest quality musical recording. When used out of doors—e.g., for recording bird songs—they are affected by wind noise (although it is possible to use a wind shield).

Some early Grundig dynamic microphones are low impedance types which have a transformer built into the lead to connect to high impedance inputs. Others—e.g., the GDM5—have no transformer and are intended for low impedance microphone inputs.

One Grundig moving coil microphone,—the Grundig GDM III, has a cardioid (heart-shaped) field. This is produced by combining two sensitive elements so that the sounds from the front add up and those from the back cancel out. If you imagine a heart shape with the microphone placed at the dent in the top of the heart, the area of good reception extends in front of the microphone like the lobe of the heart. So it acts as a directional microphone. This type is excellent for recording in front of an orchestra or stage when you want the performance but not the audience noises.

RIBBON MICROPHONES. If you are interested in really superb quality, you will have to buy a ribbon microphone. The working element is a thin ribbon of aluminium, about 1/10,000 in. thick, suspended between the poles of a powerful magnet. Naturally, an element of this type is extremely fragile and calls for a lot of care in manufacture. So a ribbon microphone is an expensive instrument and you must handle it carefully.

A ribbon microphone is more sensitive to noises from in front and behind than it is from the sides. If you draw a map of the area where it is most sensitive you get a picture like a figure eight with the microphone in the middle. You have to bear this picture in mind when you are using a ribbon microphone and be sure to get the things you want to hear within the lobes of the figure eight—e.g., if you are recording an interview, have the microphone placed so that each of the parties is in one of the areas of maximum sensitivity—not in the 'dead' zones.

Although a ribbon microphone gives better quality than most other types, it does not mean that it would sound noticeably better on your recorder. You can only really appreciate the maximum performance of a ribbon micro-

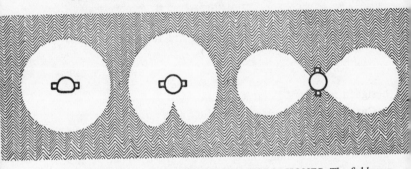

DIRECTIONAL PROPERTIES OF MICROPHONES. The field over which the signal is strongest is shown white. Omnidirectional (left) field as in most condenser, crystal and dynamic microphones. Cardioid (centre) field as in certain dynamic microphones. Figure Eight (right) field as in most ribbon microphones.

phone when you play the recording through a separate high fidelity sound system (p. 131). Unless you are going to reproduce your tapes this way, it may not be worth your while to go to the expense of a ribbon microphone.

Ribbon microphones are susceptible to damp and much more sensitive to knocks and rough handling than ordinary microphones. They are not recommended for use out of doors because they are sensitive to air currents. When even a light breeze blows on a ribbon microphone it records like a rumble of thunder.

CONDENSER MICROPHONES. This type of microphone is operated by the pressure of the sound waves acting on a thin metal diaphragm. The diaphragm forms part of an electrical condenser system, and as it moves under the varying pressure of the sound vibrations it changes the electrical capacity

148

of the circuit. These electrical changes form the signal that your Grundig amplifies and records on the tape.

The Grundig condenser microphones are equally sensitive to sounds coming from all directions (*i.e.*, they are omnidirectional) and are therefore excellent for general purpose work. The quality is well up to the standard you want for making good records of live music.

Don't try to plug a condenser microphone into a socket intended for another type. The condenser type needs a special D.C. polarising voltage of 100 volts before it will work. This is provided on the microphone input of most of the early Grundig models. Nowadays most models are designed for moving coil microphones.

CRYSTAL MICROPHONES. The working element in this type of microphone is a special crystal that turns sound vibrations into electrical impulses.

Compared to other microphones, the crystal gives a lot of electricity for a little sound, and it responds to sounds from all around it. Crystal microphones are cheap, and they are as robust as you can expect anything as sensitive as a microphone to be. But they don't like heat and damp; keep your crystal microphone in a normal working temperature and don't blow damp breath on it to see if it is working.

Crystal microphones tend to respond more strongly at the higher frequencies—around 6,000 cycles per second—and this gives a characteristic brilliance to speech and music. By careful attention to the design of the crystal element and the equalising circuit in the tape recorder the response can be made reasonably flat over the whole range of frequencies.

MICROPHONE IMPEDANCE: Some Grundig models are designed to record from high impedance types of microphone and others from the low impedance types. Broadly speaking, the high impedance types are those with the impedance value stated in thousands of ohms and the low impedance types in hundreds of ohms or less. The early Grundig models had high impedance condenser microphones and there was one high impedance crystal microphone. Some of the earlier moving coil microphones were

149

high impedance and some low. Most of the recent moving coil microphones have two outputs—a high impedance, connected across pins 1 and 2, and a low impedance, connected across pins 2 and 3.

Stereo microphones

To make stereo records with the Grundig stereo models, you need two separate microphones—one connected to each channel. Both microphones must be identical, as they are controlled by a single knob and the recording level is indicated by one magic eye. The microphones are placed symmetrically to right and left of an imaginary line coming out of the middle of the source of sound being recorded. The stereo effect will vary according to the distance separating the microphones, and you have to make a series of recordings starting with the microphones close together and then moving them away from each other until you get the effect you want. Different subjects and recording conditions will call for different microphone spacing.

The best stereo recordings are made with the special stereo microphones. These units consist of two directional microphones mounted on a common stand, one above the other. The microphones are mounted so that they can be turned to increase or decrease the angle between their directional axes. Here again, you have to run a series of tests to find the best angle, remembering that each microphone must have its axis inclined at the same angle to the centre line of the subject.

Microphone stands

All the Grundig microphones are designed to be held in the hand or, if you like you can stand them on a desk or shelf.

If you are making records in the garden, then stand the microphone on a wall or step, or bring a stool out of the house and put it on that. But whatever you use, don't have the microphone on or very close to the recorder itself, or you may pick up electrical hum and mechanical noise from the motor.

When you hold the microphone in your hand you must take care or you will pick up a lot of sounds from handling that you don't want. The slight scraping of the microphone lead on the floor can also record a noise on the tape, and every move you make will sound like the rumblings of distant thunder. So, if you must hold the microphone, try it out first, and then you'll learn how to handle it to keep these unwanted noises down to a minimum.

For stereo recording, the microphones must not move.

When you record monaural sound with an omni-directional microphone, it doesn't matter which way the microphone is pointing, and if you are carrying it in your hand, you can move around as much as you please. But when you are recording stereo sound, the microphones must be fixed in one position and left there throughout the recording. So the stereo microphone unit is always mounted on a stand of the desk or floor type.

Mixers

A mixer is used for mixing together two or more signals and feeding them through a single lead into your Grundig. It permits you to mix signals coming from sources of widely varied characteristics, and to balance one with the other so that they blend just as you want them. Alternatively, you can alter the emphasis on any particular signal as you record. With some you can monitor the final result on head-phones direct from the mixer; otherwise you can monitor with headphones plugged into the recorder output.

GRUNDIG MIXERS, GMU 1 and GMU 2, now obsolete, are simply resistive networks—*i.e.*, they have no power to boost any particular input; they can only level *down* one that is too powerful and cannot lift up a weak signal.

GRUNDIG MIXER GMU 3 combines a resistive network and a valve amplifier so that it can actually amplify the different signals as well as cut down their strength. This mixer runs off the 200–250V. A.C. mains supply.

THE GRUNDIG STEREO MIXER is a transistorised unit powered by two Type PP3 dry cells. It can handle inputs from a stereo microphone and a stereo radio or stereo gramophone. It can also mix a monaural signal with the stereo channels and has a separate control to fade a monaural signal into either stereophonic channel at will. This model has sliding, studio-type controls in place of the usual rotary controls.

The Grundig Channel Reproducer CR1

This is a self-contained amplifier/speaker unit designed to give improved reproduction quality and volume when you plug it in to the high impedance output of models fitted with a switch for cutting out the built-in speaker. It can also be used for reproducing the second channel of a stereo tape played on the TK24 and later 4-track models (p. 34). The Channel Reproducer has its own volume and tone controls—the volume and tone controls on the tape recorder have no effect on the high impedance output. When you play your tape through the Channel Reproducer you will usually cut out the built-in speaker with the speaker switch.

The Grundig remote controls

Some Grundig models are designed to be stopped and started by remote control and the two-way deck models can be stopped, started and back-spaced by remote control. This has a number of advantages—especially for dictation.

All the Grundig remote controls plug into a special socket on the back panel of the recorder. Some models have a socket to take a DIN type five-pin-plug; the remaining models use a concentric plug with five metal ribs.

The actual control is a metal, foot operated switch with a flexible lead. There is a hand operated press-button model for stop/start only.

When you plug in one of these units and set the main controls to record, you can stop, start and, with certain models (on playback only) reverse the machine by regulating your foot pressure on the control pedal.

When your typist comes to transcribe the record on to her typewriter, she uses the same set-up and simply plugs a pair of stethoscope earphones (below) into the appropriate socket on the back panel of the recorder. She then switches the main controls to playback and then uses the foot control to play back or backspace the tape to play again anything she wants to check as she goes along. All the above operations apply to both tracks on two-way deck machines, depending on which track key is pressed. The normal fast wind keys can still be operated by hand, and the automatic stop comes into action as usual at the ends of the tape.

The control has a number of other uses—*e.g.*, in editing the tape, when you want to hit the exact spot where a word begins or ends, for repeating foreign language pronunciation, and so on. A remote control is also useful when you want to make candid records (p. 69). You can hide your Grundig away in a cupboard and put the foot switch of the remote control under the table where you are sitting. With this arrangement you are free to switch on and record the conversation whenever you like.

There are eight versions of the remote control and they are not necessarily interchangeable between one model and another. The data section at the back of the book tells you which controls are suitable for which Grundig models.

The Grundig Stethoscope earphones

You use stethoscope earphones when you want to listen to your Grundig with the speaker turned off or for transcribing a recording, or listening in privacy.

The unit consists of a miniature dynamic receiver mounted at the hinge of a pair of plastic tubes curved like a stethoscope to fit into the ears. The earphones are extremely light and are in no way irksome to wear for long periods. From the woman's angle they are welcome because there are no projections to catch in the hair. There are two versions of the stethoscope earphone. The correct one for your Grundig

153

model is given in the data section. You can buy the receiver fitted with a single ear clip to which you can add the Stethoscope Attachment for two-ear listening.

The Grundig telephone adaptor

A Grundig telephone adaptor allows you to make a tape recording of both sides of a telephone conversation without interfering with the wiring of the telephone in any way. It consists of a coil, in a plastic case, with a rubber sucker attachment by which you can stick it on the side of the telephone base. The coil is fitted with a screened, flexible lead and plug to fit the diode (or telephone adaptor) input socket on the back panel of your tape recorder.

To get the best results you have to experiment with the position of the adaptor on the telephone base. As the coil is sensitive to all magnetic fields, including the one set up by domestic electric wiring, you may also have to try moving the telephone and tape recorder about the room to get rid of mains hum.

There are two Grundig telephone adaptors, each intended for use with certain Grundig models only. One is terminated with a jack plug, the other with a three-pin plug.

Grundig spare leads

The Grundig spare leads are designed for connecting external equipment to the various models—*e.g.*, connecting radio sets, gramophone pick-ups, mixing units, etc., to the input sockets and connecting external speakers, amplifiers and other tape recorders to the output sockets. These leads are made up with suitable plugs at one or both ends. A complete list is given under Accessories in the Data Section.

Grundig microphone extension leads

In addition to the above spare leads you can buy made-up screened leads in lengths from 5 to 18 yds. for extending the normal Grundig microphone leads. These leads are available for all the microphone types—condenser, crystal.

154

dynamic and ribbon—but it is essential to use the correct type in each case; the leads are not always interchangeable and you should ask your Grundig dealer for the correct type for your particular microphone. The wrong type of lead can seriously distort the signal.

You can extend the lead of the GCM 1 microphone up to 15 yds. without affecting the frequency response, but you must use low capacity closely braided co-axial cable. Grundig extension cables MC 5, 10 or 15 (the figures indicate the lengths in yards) are best for the purpose. With the longest lead the sensitivity of the microphone will drop by about 4 decibels (dB). (If you have some knowledge of electronics and want to make up your own leads, you should choose a cable with a capacity of not more than 20 pf/foot; Telcon cables PT 1 M and K 19 M are suitable).

The lead to a GCM 3 microphone can also be safely extended to 15 yds. The correct Grundig extension cables to use with this microphone are MEC 5, 10, or 15.

The GDM 5Z and GRM 1/2Z microphones for model TK12 are low impedance types and have matching transformers built into the lead to match them to the high impedance microphone input. The matching transformer must be kept close to the tape recorder, so when you extend the lead, the extensions must be inserted between the microphone and the transformer. The correct way to extend these microphone leads is with one of the Grundig spare leads (above). The microphone lead should then be broken between the microphone and the transformer and the ends fitted with a suitable plug and screened socket to correspond with the extension lead. You can then insert or remove the extra length when necessary.

Grundig spare plugs and sockets

Most Grundig tape recorders are fitted with standard DIN-type input and output sockets. The early models have sockets for 6 mm. jack plugs. Spare plugs and sockets are available as listed under Accessories in the Data Section to let you connect other equipment to your Grundig.

Grundig Transistor Model Mains Packs

The Grundig Mains Packs let you run transistor models off the ordinary A.C. power supply.

Grundig Monitor Amplifiers

The Grundig Monitor Amplifiers are designed for use with the four-track recorders in conjunction with a Grundig Stethoscope Earphone. They let you listen to Track 1 while recording on Track 3, or Track 2 while recording on Track 4. MA1 is for the TK24; MA2 is for later models.

When you have recorded the first track, you wind back and then plug the amplifier into the output socket of the recorder and the earphone into the amplifier socket. As you record the second track, you hear the first record in the earphone so you can synchronise the two records. You can then play back both tracks together (p. 38).

Grundig Distributor Speaker

The Grundig Distributor Speaker is designed to impart extra crispness and brilliance to the upper frequency reproduction of the TK830/3D. It consists of two speakers in a special type of mounting that is designed to overcome the tendency of high frequencies to radiate in a beam.

Grundig Sonodia Slide Change Adaptor

This accessory enables you to use your Grundig to play back a commentary on a series of slides and operate the slide changing mechanism automatically by means of impulses recorded at the appropriate places on the tape. See slide shows, P.114.

The Grundig Stereo Mixer

The stereo mixer is designed to mix 4 inputs—e.g. from a microphone, radio, gramophone and second tape recorder. There are 3 coarse level controls, a directional control and control for a reverberation unit. It has 4 microphone pre-amplifiers and may be used for mixing two stereo or 4 mono signal sources. Power is supplied by two 9 volt batteries.

CHAPTER 10

Keeping Everything Right

TO GET THE MAXIMUM USE AND ENJOYMENT out of your Grundig you must take care of it and give it proper attention from time to time.

Looking after your Grundig

You can expect your Grundig to run for several hundred hours—*i.e.*, from six months to a year of normal use—without any attention at all. All the important bearings are self-lubricating and there is practically nothing to wear in the control mechanism because many of the operations are carried out by electrical relays—the press keys simply make or break electrical circuits.

About the only maintenance job you need trouble about is to keep the heads clean. As the tape runs through the sound channel and passes over the heads, a certain amount of the magnetic coating gets rubbed off. This tends to build up and prevent close contact between the tape and the face of the head. The result is loss of quality first on recording and then on playing back the tape.

It is a simple matter to prevent this by cleaning the heads from time to time. First you have to take off the sound channel cover. On some models you do this by squeezing the ends on the front half of the cover inwards and lifting it off; on other models you have to lift off the whole deck—first removing the control knobs and then undoing the screws holding the deck.

With the heads exposed you will be able to see the brown deposit of oxide at each side of the polished face of the erase and record/playback heads. It will clean off readily whit a piece of old cotton handkerchief or similar non-fluffy material wrapped around an orange stick. Any stubborn deposit can be softened by moistening the rag with methy-

CLEANING HEADS. To clean heads, lift off sound channel cover and remove brown oxide deposit from polished face of head with dry camel-hair brush or soft cloth around an orange stick.

lated spirit. While you are at it, wipe off any oxide accumulation in the sound channel and on the tape guides.

There should be no need to touch the capstan and pressure rollers, but if they need cleaning (p. 160) a piece of fluffless rag moistened in methylated spirit will do the trick.

You can buy spools of special cleaning tape that you can run through the sound channel to clean both the heads and the capstan without removing the sound channel cover.

To keep your Grundig spick and span give it an occasional rub over with anti-static polish. This stops the surface from getting electrically charged and attracting dust.

This amount of attention should keep your Grundig in first class working order for at least two years. At the end of that time you should get your Grundig dealer to check the adjustments and see that everything is in order. With his special knowledge and test equipment he can do the job in a fraction of the time you would take. The finest equipment can go out of adjustment after long periods of use and the resulting loss of quality takes place so slowly that you may not notice it. A periodical check up with the proper instruments will make sure that you get the best out of your Grundig.

In the course of time (and sometimes after an electrical fault) the record/playback head may become permanently magnetized. When this happens the recording and reproduction quality will suffer and you will notice an unpleasant amount of background 'mush'. There is a special tool that will remove all traces of permanent magnetism in a few seconds. You don't need to buy one; your Grundig dealer almost certainly has one and will demagnetize your sound heads *in situ* for a nominal sum. It is always worth having this done about once a year.

Troubles and how to cure them

The faults listed on the next page are the only ones that you can safely do anything about.

If you encounter any trouble not dealt with below, hand the job over to your Grundig dealer. All Grundig dealers have the experience and the necessary special tools to trace and cure faults quickly.

After a lot of use, it is a wise plan to have your Grundig inspected even if it has been giving good service anyway. Small defects that build up over time easily pass unnoticed if you have been 'living' with them; but the experience of your dealer and the test equipment that he has can quickly tell you if you are, in fact, still getting the best results.

Trouble	Cause and Remedy
Motor runs but spools do not turn.	(1) Tape threaded incorrectly. Press stop key, pull tape taut and check that it is free in sound channel (p. 18). (2) You have forgotten to wind foil insert through the sound channel.
Tape jumps out of tape guides.	Dirt or oxide coating on pressure roller or capstan. Clean (p. 158).
Tape scraping against edge of spool.	Spool is warped. Replace with new one.
Tape does not run at correct speed.	Tape not threaded correctly. Check threading (p. 44).
Tape breaks on starting fast wind.	You have allowed slack tape to form between the spools. Always check that tape is taut at both ends of sound channel before starting.
Tape breaks during record or play-back.	(1) Tape is incorrectly threaded. (2) Tape has been spliced with unsuitable jointing tape (p. 93) and is sticking to rollers in sound channel.
Reproduction sounds woolly and lacks top response.	Dirt or oxide on face of record/playback heads. Clean (p. 18).
New recording is marred by traces of previous recording on same tape.	Incomplete erasure of previous signal. May be due to recording at excessive signal level. Run through sound channel as for recording but with recording level turned right down and no input connected.
Background noise level is unpleasantly high.	Original recording level was much too low (p. 45).
Unsteady, wavering quality noticeable on sustained notes (called Wow)	Dirt, oxide or oil on capstan or pressure roller. Clean (p. 18).
Excessive hum when playing back.	(1) Mains interference. Try reversing 2-pin mains lead connexion. (2) Input connexions are faulty (p. 139).
Weak or no signal even with signal level control full on.	Input plug in wrong socket on back panel (or mixer panel). Change to correct socket.

Glossary of Tape Terms

AMPLIFIER. A piece of equipment that increases the strength of a weak electronic signal.

AMPLIFYING UNIT. The part of a tape recorder that includes the distribution panel, bias/erase oscillator, record and playback amplifier, equalising circuits, tone correction and all associated controls.

A.M. (RADIO). Abbreviation of 'Amplitude Modulated'. This type of radio is more subject to atmospherics and interference than Frequency Modulated Transmission.

AUDIO FREQUENCY. The frequency (speed of vibration) of a sound wave that can be heard by the human ear. The range of these audible waves lies between 30 and 15,000 cycles per cecond.

AUTOMATIC STOP. Mechanism incorporated in a tape recorder to disconnect the tape transport mechanism automatically when all the tape has run off the spool.

BAFFLE. Flat rigid surface in which a speaker is mounted to increase its efficiency.

BATTERY. Self-contained source of electric power consisting of either re-chargeable cells or cells which actually generate electricity by some type of chemical action.

BULK ERASER. A device for erasing recorded signals from the whole of a tape or a number of tapes at once. It generates an alternating magnetic field in the same way as the erase head on a tape recorder, but of much higher intensity. It does the job in a second or two and avoids the lengthy process of erasing by running through the sound channel.

CAPSTAN. One of the pair of pressure rollers which grips the tape and draws it through the sound channel at a constant speed. The capstan roller is driven by a motor and is the one that actually moves the tape; the other turns freely on its spindle.

CASSETTE, TAPE. Flat container holding two tape spools permanently threaded and ready to record or play back.

C.C.I.R. Stands for Comité Consultatif International des Radio-communications. This international body has published a recommended specification for recording characteristics.

CELLULOSE ACETATE. Plastic made from cellulose and used as the flexible base for some types of magnetic recording tape.

CONVERTER. Piece of electrical or electronic equipment which converts direct into alternating current.

COPYRIGHT. The right of the author of a creative work to enjoy the rewards of his efforts, in particular the right of a musician to the proceeds of the work he has composed or recorded as defined by the Copyright Act.

CROSSOVER FILTER. An electrical filter circuit which divides up the output of an amplifier into two or more frequency bands. Each band of frequencies is then fed into a separate speaker specially designed to handle that particular band. The results are generally superior to those produced when a single speaker has to handle the whole range of frequencies.

CYCLE. A musical note and a radio signal are the result of regular vibrations resembling waves. Every wave passes through the same 'life history'—it starts at zero, rises to a maximum value, falls to a minimum value and then starts all over again. The sequence of events from beginning to end is called a cycle.

DECIBEL (dB). Unit commonly used for measuring sound intensity. By coincidence, one decibel is the smallest change in the volume of a sound that can be detected by the human ear.

DISTORTION. Discrepancy between the sound reproduced by a piece of equipment—*e.g.*, an amplifier—and the original signal.

DISTRIBUTION PANEL. The panel in a tape recorder carrying sockets or terminals for connecting to external equipment—*e.g.*, mains, microphone, amplifiers, speakers.

DUBBING. Transferring a recording or parts of it from one tape to another by using two tape recorders.

DYNAMIC RANGE. The ratio between the softest and loudest signals (measured in decibels) that sound equipment can reproduce without distortion.

EDITING. Technique of improving original recordings by the removal of unwanted sections, the insertion of new material, and the introduction of special effects.

ELECTRO-MAGNET. A coil of wire exhibiting the properties of a magnet when a current of electricity is passed through it. The presence of magnetic material—*e.g.*, iron—in the centre of the coil greatly increases the effect.

ELECTRONIC. Associated with the controlled flow of electrons as in thermionic valves, cathode ray tubes and transistors—*e.g.*, amplifiers, radios, T.V. sets.

EQUALISATION. Automatic electronic correction applied to a signal before or after recording to compensate for disproportionate amplification of upper and lower frequencies. Pre-amplifiers provide separate equalising circuits to deal with various types of gramophone pick-up; a tape recorder circuit is designed to boost the higher frequencies in a recorded signal and suppress them during playback.

ERASE. To remove the recorded signal from a magnetic tape by exposing it to a powerful magnetic field—usually one alternating at high frequency.

ERASE CUT-OUT. A switching arrangement which disconnects the erase head so that a second signal can be recorded on the tape without wiping out the first.

ERASE HEAD. The first head that the tape passes over in the sound channel of a tape recorder and which automatically erases any previous signal when the instrument is set to record.

EXTENSION SPEAKER. An independent speaker connected to a tape recorder (or radio set) either to relay the signal to some other place or to substitute a larger and better speaker for the one built into the tape recorder.

EXTERNAL AMPLIFIER. An amplifier connected to the output of a tape recorder to increase the power or quality or both of the signal. Generally, when an external amplifier is plugged into the tape recorder it takes the entire output and the built-in speaker is cut out.

FEEDBACK (ACOUSTIC). The build-up of the signal that takes place when a microphone is connected to an amplifier and speaker in the same room. Sound waves from the speaker enter the microphone and are amplified. They emerge from the speaker louder than before and 'feed back' into the microphone. The result is a loud, continuous squeal from the speaker.

FEEDBACK (NEGATIVE). An electronic trick for reducing distortion. Undesirable frequencies in the output stage of the amplifier are looped back to an earlier part of the circuit and fed in, as it were, upside down. The amount of feedback is adjusted until it exactly cancels out the distortion. Negative feedback is also used to 'throttle down' the amplifier when there is only a small load on the output stage.

FLUTTER. A rapid wavering in the sound from the speaker when playing back a tape recording, often caused by some irregularity in the tape drive—*e.g.*, a shred of tape stuck to the capstan roller, or a spot of grease on the friction surface.

F.M. (RADIO). Abbreviation of 'Frequency Modulated'—a method of transmitting a radio signal by causing it to vary the frequency of the carrier wave. This type of transmission is practically free from interference from other stations or from atmospherics.

FREQUENCY. Applied to an alternating current or electromagnetic field means the number of complete cycles occurring every second.

FREQUENCY RESPONSE. Measures the ability of a piece of sound equipment to reproduce signals varying from the lowest to the highest frequencies that are audible to the human ear. This is commonly given in the form of a graph showing the relationship between input and output volumes from about 40 cycles per second to 15,000 cycles or more.

If the circuit amplifies all frequencies in the same proportion, the graph takes the form of a straight horizontal line and the equipment is said to have a 'flat response'. In practice this statement is qualified by some such phrase as 'plus or minus so many decibels'. A frequency response of 15 to 30,000 c.p.s. plus or minus 1 decibel or 15 to 50,000 c.p.s. plus or minus 2 decibels is up to good high fidelity standards.

GAIN. The 'amplification factor' of an amplifier—*i.e.*, the number of times the volume of the output signal exceeds that of the input.

GAP. The space separating the poles of a tape recorder head at the point where the tape passes over it. The effect of the gap is to produce a concentration of magnetic flux in the magnetic coating of the tape. There may be a second gap on the opposite side of the ring formed by the two halves of the core, known as the back gap.

HEAD. One of the electromagnets in the sound channel of a tape recorder which either erases, records, or plays back the signal.

HIDE. A screen used by bird and wild-life observers or recorders to let them get close to the subject.

HI-FI. Slang for high fidelity.

HIGH FIDELITY. A term used to describe sound reproduction which is a completely faithful representation of the original sound over the whole range of audible frequencies.

H.T. Abbreviation of 'High Tension'. Term applied generally to the anode power supply in an electronic circuit.

HUM. A steady noise of low to medium pitch present in the reproduced signal as a result either of a definite fault or simply poor design or construction. Most hum comes from the A.C. main supply.

HUMDINGER. An adjustable resistance bridge connected across the valve heater circuit in such a way that it cancels out any hum present in the amplifier output by injecting an equal and opposite amount of 'artificial' hum.

IDLER. Alternative name for pressure roller.

IMPEDANCE. Measure of the opposition offered by a circuit to the passage of an alternating current. It is measured in ohms and is the combined effect of resistance to *flow* as encountered in direct current circuits and resistance to *change* in strength (known as *reactance*) which is peculiar to fluctuating currents. Impedance is important when connecting two pieces of electronic equipment such as a tape recorder output to an external amplifier input.

INPUT. The signal fed into a piece of electronic equipment—*e.g.*, a tape recorder or an amplifier.

I.P.S. Abbreviation of 'Inches Per Second'. Unit for measuring the speed of magnetic tape through the sound channel when recording or playing back.

JACK. A socket fitted with internal spring contacts which enables two electrical circuits—*e.g.*, a tape recorder and an external amplifier —to be connected via a fitting plug. This arrangement provides a quick and foolproof method of connecting pieces of equipment.

LEVEL. The average strength at which the input signal is maintained when making a tape recording. It should be such that only occasional intense volume peaks rise above the value at which distortion is known to occur.

LEVEL INDICATOR. An instrument (microammeter or magic eye) mounted on a tape recorder which gives a visible indication of the strength of the signal being recorded on the tape. It helps the user to adjust the level control to the highest level short of the point where distortion begins.

LOUDSPEAKER (or simply **SPEAKER**). Device for converting an electrical signal into sound.

L.T. Abbreviation of 'Low Tension'. Term applied generally to the heater current supply in an electronic circuit.

MAGIC EYE. A visual indicator used as a guide to help you to adjust the signal level. It is formed by a pattern of fluorescence in a special type of valve. The fluorescence usually takes the shape of two luminous lines or segments. When the two parts just touch and fill the whole of the 'eye', the signal is at the maximum level short of distortion.

MAGNETIC TAPE. The sensitive material on which the tape recording is made. Made from a flexible base coated with a magnetic iron oxide.

MAGNETISM. Power of attraction found in 'magnetic' iron ore which can also be produced by passing an electric current around a coil of wire.

MICROPHONE. Any device that converts sound waves into electrical signals. There are five principal types—carbon, moving coil, crystal, ribbon and condenser.

MIXER. A tape recorder accessory which enables signals from several different sources of different impedances and levels—*e.g.*, radio, pick-up, microphone—to be separately controlled and then mixed in any proportion to give a single signal at the required level.

MODULATION. Transmitting or recording a signal by imposing it on a uniform medium—*e.g.*, a carrier wave as in radio communication.

MONITOR. To keep a running check on a signal that is being recorded or transmitted.

MOTOR BOARD. The panel on the tape recorder that carries the tape driving equipment with its motor or motors.

NEGATIVE FEEDBACK. See Feedback.

NOISE LEVEL. The level of unwanted noise—*e.g.*, hum, tape hiss, background noise—present in a recorded signal in relation to the level of the signal alone.

OSCILLATOR. The part of a tape recorder electronic circuit that supplies the high frequency alternating current used for (a) erasing the tape and (b) biasing the signal being recorded.

OUTPUT. The power delivered by sound reproducing equipment to the speaker. It is measured in watts.

PATCH CORD (U.S.A.). A lead fitted with suitable plugs or other connectors for joining two pieces of electrical equipment.

PLAYBACK. The operation of reproducing a tape record.

PLAYBACK AMPLIFIER. The part of a tape recorder amplifier unit used for amplifying the signal 'read' off the tape by the playback head, applying the necessary equalising compensation and feeding it to the speaker.

PLAYBACK HEAD. The head in the sound channel of a tape recorder responsible for converting the magnetic pattern recorded on the tape into an electrical signal which can be amplified and reproduced through the speaker. The playback head is usually situated last in the sound channel. On many recorders the signal plays back and records with the same head.

PLUG. A fitting that connects the conductors in a flexible lead to the appropriate contacts in a socket to form a continuous electrical circuit between two pieces of equipment.

POLYESTER BASE. Flexible base of a magnetic tape made from polyester instead of acetate film. This is stronger and stretches less than acetate. It is used for the thinner 'long play' tapes and for carrying sound tracks for synchronising with cine films where stretch in the tape would put the sound out of step with the picture.

POLYVINYL CHLORIDE (P.V.C.). A tough plastic used as a base for magnetic recording tapes.

POWER AMPLIFIER. The amplifier that steps up the weak signal from the playback head to the level required to operate the speaker.

PRE-AMPLIFIER. A piece of electronic equipment equipped with circuits which will equalise, amplify or weaken (attenuate) input signals to produce a signal with characteristics suited to the amplifier or tape recorder. Most tape recorders and some amplifiers incorporate pre-amplifier circuits capable of handling the more important inputs—*e.g.*, radio tuner, gramophone pick-up and high and low impedance microphones.

PRE-RECORDED TAPES. Commercially recorded magnetic tapes sold by certain recording companies and serving the same purpose as ordinary disc records.

PRESSURE PADS. Spring-loaded felt pads which hold the magnetic tape in close contact with the heads in the sound channel of a tape recorder.

PRESSURE ROLLER (or IDLER, PINCH WHEEL). One of the pair of rollers that grip the magnetic tape and draw it through the sound channel at a constant speed when recording or playing back. The pressure roller is free to rotate and is simply there to back up the pressure exerted by the motor-driven capstan.

PRINT THROUGH. Transfer of the more heavily recorded parts of a signal to the adjacent layers of tape on the spool. The effect is more pronounced at high temperatures and with the thinner grades of tape. It is unlikely to happen if the recording level is kept below the distortion point and the tape is stored in normal room temperature.

PRINTED CIRCUIT. A radio or tape recorder circuit in which the normal wiring is replaced by metal lines printed or plated on to an insulated backing. This type of wiring is more robust and is particularly suitable for portable equipment.

P.V.C. Abbreviation of Polyvinyl Chloride.

RAW TAPE. Tape before recording.

RECORD (TAPE). A magnetic tape carrying a magnetic recording made by sounds which are reproduced more or less faithfully when the tape is played back on a tape recorder.

RECORDING AMPLIFIER. Part of the tape recorder amplifying unit used for amplifying the signal to be recorded on the tape.

RECORDING HEAD. The head in the sound channel of a tape recorder that converts the electrical signal to be recorded into a magnetic pattern on the tape, the pattern forming a record which can be reproduced by playing back the tape. On many recorders the recording and playback heads are one and the same; on others the recording head is separate. Where there is a separate recording head, it comes before the playback head in the sound channel.

RECORDING LEVEL. The level to which an input signal is controlled so that only very occasional peaks rise above the point where 'overloading' of the tape and consequent distortion of the recording begin.

RECORDING LEVEL INDICATOR. A visual indication of the signal level; generally a magic eye or meter.

REFLEX CABINET. A popular form of speaker mounting in which the sound radiation from the back of the cone passes around an enclosure and out of a port, emerging in phase with the direct radiation from the front of the cone.

RELAY. An electrical switching device in which a comparatively feeble current is made to release a more powerful one.

REMOTE CONTROL. An accessory available with some tape recorders that allows them to be operated from a distance through a flexible lead with a separate control unit at the end.

RESIDUAL NOISE. Noise remaining on the tape after it has passed over the erase head and appearing as a background to the new recording.

167

REVERBERATION. The more or less gradual dying away of sounds in a room caused by successive reflection of the original sound from the walls, floor and ceiling.

RPM. Abbreviation of Revolutions Per Minute; a measure of speed of rotation.

SCREENED CONNECTING LEAD. Lead for connecting two pieces of electronic equipment consisting of one or more insulated conductors inside a metal braid sheath, The sheath is connected to earth and acts as a screen which prevents the inner conductors from picking up electrical interference.

SELF-POWERED RECORDER (U.S.A.). A battery operated portable model.

SIGNAL. Electrical oscillations received from microphone, pick-up radio, etc., which record on the tape or which are generated by the playback head, amplified and reproduced by the speaker.

SIGNAL-TO-NOISE RATIO. The ratio of the strength of the loudest undistorted signal that the tape can reproduce to the strength of the background noise is the signal-to-noise ratio.

SINGLE (FULL) TRACK RECORDER. Some early tape recorders and a few modern machines for extra high quality reproduction record a single track down the centre of the tape. Most other tape recorders make two records side by side on one tape (half track recorders) or four records side by side (quarter track recorders).

SOLENOID. A coil in which slides an iron core. When current is passed through the coil the core is pulled into the centre with sufficient force to operate mechanisms like switches, tape transport controls and so on.

SOUND. Audible vibrations, usually in the air, set up by any form of rapid cyclic movement having a frequency lying between 20 and 20,000 vibrations per second. Strictly speaking, vibrations of a higher frequency are not sound, although they are often referred to as sound, and in fact, although inaudible to the human ear, can be perceived by various animals and insects.

SOUND CHANNEL. The part of the tape deck between the spools which houses the erase, record, playback (or record/playback) heads and the capstan assembly.

SOUND HEAD. General term for record/playback heads.

SPEED CONTROL. Lever or knob on the tape deck that gives a choice of two or sometimes three tape speeds.

SPLICE. A join in a length of magnetic tape formed either by butting the ends together and fixing them with adhesive jointing tape or by overlapping the ends and welding them together with jointing compound.

SPLICER (SPLICING BLOCK, SPLICING JIG). An accessory for making the job of splicing easier and quicker. The tape ends are overlapped and held in a channel which carries a guide for a cutting knife, and prevents the tapes from slipping out of register while they are being spliced or welded.

SPLICING TAPE. Special self-adhesive tape made for splicing magnetic tape. The adhesive does not tend to ooze out at the sides of the joint and cause sticking.

SPOOL. The flat bobbin on which magnetic tape is sold and loaded on to the tape deck.

STEREO. Term for sound recorded and reproduced on two channels to create sensation of natural hearing.

SYNCHRONISM. The condition of being 'in step', as when a sound recording is played back to coincide exactly with the corresponding sequence of events in a projected cine film.

TAPE DECK. The part of a tape recorder which takes care of the physical operations of winding the magnetic tape from supply to take-up spool through the sound channel and back again.

TAPE GUIDES. Posts of non-magnetic metal placed at each end of the sound channel so that the tape enters the sound channel at the same angle irrespective of the amount of tape on the reel.

TELEPHONE ATTACHMENT. An electro-magnetic pick-up which can be attached to the side of a telephone instrument and connected to a tape recorder to record both sides of the conversation.

TEMPORARY STOP. The control used for short pauses during recording or playback. It disengages the capstan and pressure wheels, leaving the capstan running but out of contact with the tape. The normal stop/start controls are not involved.

TEST TAPE. A magnetic tape pre-recorded with a range of continuous tones and other items which enable the performance of a tape recorder to be checked by actual trial. Test tapes are listed in most pre-recorded tape catalogues.

THREADING SLOT. The slit along the length of the sound channel cover which allows the tape to be slipped into position in the open channel.

TONE CONTROLS. Controls on the tape recorder amplifier to adjust the balance of frequencies on playback. There may be separate controls for lifting or cutting bass and treble independently, or there may be only one control giving treble cut. By designing the circuit to give excessive treble with the control at maximum, a single control can be made to give a range of tones from shrill to 'boomy'. Single control is only used on the less expensive popular models.

TRACK. The actual area on a magnetic tape that carries the recording. A tape may be recorded on one, two, four or even more tracks.

TRANSISTOR. A current multiplying device which can take the place of valves in electronic equipment and which generally operates at a lower voltage and consumes only a fraction of the electricity required

M 169

by a valve—hence its use to replace valves in battery-operated portable tape recorders.

TWEETER. Slang term for the small speaker used in conjunction with the normal speaker to improve high frequency response.

VOLUME CONTROL. A continuously variable resistance inserted in a circuit—generally associated with either input or output—to give a progressive increase or decrease in signal strength. The resistance is designed in such a way that movement of the control produces a roughly proportional change in single strength.

WELDING. Joining magnetic tape by the use of a liquid cement.

WOOFER. Slang term for a large diameter speaker devoted exclusively to the reproduction of the lower frequencies—*i.e.*, 30 to 800 c.p.s.

WOW. Fluctuations in the pitch or volume of a signal when played back. Wow is similar to flutter (above), but it is slower and comes from other causes.

Data Section:
All the Grundig Details

In the following pages you will find all you are likely to want to know about all the Grundig models and the accessories you can use with them.

Working Instructions—in a nutshell are given for recent models only.

The individual model data sheets group several models together where the layout of controls is similar, but the important differences between the models in each group are given in each case.

If you want to connect non-Grundig equipment to your machine, always make sure that the impedance is suitable. With most of the equipment you are likely to buy, there is a margin of latitude, and so long as you specify 'high' or 'low' impedance, you won't go wrong. However, to be on the safe side you should always check your choice of equipment with your Grundig dealer.

Important

Your Grundig is designed for 50 c/s A.C. mains. You will damage it seriously if you connect it straight to a D.C. supply. If you want to run it off a D.C. supply, you will need a special converter (p. 62) and should see your dealer about it.

It is dangerous to connect your Grundig to the output of an A.C./D.C. radio or of a TV set unless you make the connexion through an isolating transformer. Be on the safe side and ask your Grundig dealer to look after this job for you.

Grundig Models 700L, 700C, 500L Mk. I/II

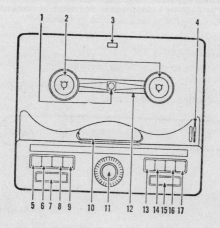

Deck Key

1 Pilot Light.
2 Spool Holders (7 in. spools)
3 Tone Control.
4 Speed Control (3¾ and 7½ i.p.s.).
5 Prepare to Play Back.
6 Diode input selector.
7 Pause Bar.
8 Microphone input selector (Grundig condenser microphone).
9 Radio L.S./Gram. P.U. input selector.
10 Sound Channel.

11 Magic Eye/On-Off Switch/Signal Level Control (Recording) and Volume Control (Playback).
12 Tape/Time Scale.
13 Play Back Gram P.U. through built-in speaker.
14 Fast Wind Right to Left.
15 Stop Bar.
16 Fast Wind Left to Right.
17 Start Tape (after pressing keys 5, 6, 8, or 9).

Model Differences

700C has an output of 4·5 watts (other models 2·5 watts).
500L Mk. I/II has only a single speed—7½ i.p.s.

Accessories

Microphones: GCM1.
Telephone Adaptor: TA1.
Stethoscope Earphones: STET 1.
Remote Control: RCF1; RCH1.
Plugs: Grundig Jack Plug (6 mm. dia.) type J1.
Connecting Leads: Screened cable terminating in J1 plug.

Tape Position Indicator

These models are not fitted with a mechanical digital counter but have a graduated scale marked on the deck between the spool holders. This scale reads directly in minutes the time the tape has been running and the time still to run at either tape speed. The figures refer to the playing time of Grundig Standard tape and must be increased by 50% when using Grundig Polyester Long Playing tape.

Mains Voltage Adjustment

1 Remove fixing screws between spool holders.
2 Lift out Magic Eye dial.
3 Lift off deck moulding to reveal Mains Voltage Selector.
4 Unscrew milled nut and set to correct value.

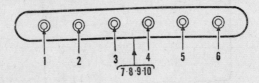

Distribution Panel Key

1 Socket for Remote Control RCH1 or RCF1.
2 Low Impedance Output (2·5 watts) for Extension Speaker or STET 1.
3 High Impedance Output (800 mV.) for monitoring or external amplifier.
4 Input for Radio L.S./Gram. P.U.
5 Input for Diode, TA1, High Impedance Microphone or GMU3.
6 Input for Grundig Condenser Microphone GCM1. (Blocking condenser must be used for other types.)

Below deck:
7 Mains Voltage Adjuster.
8 Fuses, main: 210-250, 1 amp.; 105-130, 2 amp. (Fuse not in use is stored in compartment at side of cabinet.)
9 Fuse, H.T.: 120 mA., surge resisting.
10 All sockets are designed to take Grundig 6 mm. jack plug, J1

Grundig Models: TK9, TK10, TK12, TK15, TK819, TK819A, TK820, TK919

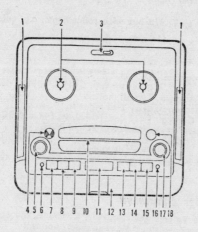

Deck Key

1 Spare Spool Compartment (TK9 only).
2 Spool Holders (5¾ in. spools on TK9 and TK12. All others 7 in.).
3 Speed Control (all models except TK9 which has only one speed).
4 Magic Eye Signal Level Indicator.
5 ON-OFF Switch. Signal Level Control (Record)/Volume Control (Play Back).
6 Record Safety Button.
7 Recording Press key (Button 1 must be depressed first).
8 Play Back Press key.

9 Track 2 Press key.
10 Sound Channel.
11 Stop Bar (Permanent. Returns all press keys to neutral. Must be pressed after last winding before any other operation is started).
12 Front Flap for spare leads on some models).
13 Track 1 Press key.
14 Fast Wind ←.
15 Fast Wind →.
16 Pause Button.
17 Tone Control/L.S. Cut out.
18 Tape Position Indicator.

Model Differences

TK9. Single speed machine, 3¾ i.p.s., 5¾ in. spools. Low impedance microphone. Single (low impedance) output. One speaker. 2½ watts output.

TK12. Two speeds, 3¾ and 7½ i.p.s·; 5¾ in. spools. High impedance microphone. High and low (2½ watts) impedance outputs.

TK10. Similar to TK12 but with two speeds, $1\frac{7}{8}$ and $3\frac{3}{4}$ i.p.s.

TK15. Similar to TK10 but with slight circuit changes.

TK819. Two speeds, $3\frac{3}{4}$ and $7\frac{1}{2}$ i.p.s., 7 in. spools. Low impedance microphone. High and low ($4\frac{1}{2}$ watts) impedance outputs.

TK819A. Chassis-only version of TK819. Sockets and wiring for connection to Grundig Arundel 8055 Console Concertgram.

TK820. Two speeds, $3\frac{3}{4}$ and $7\frac{1}{2}$ i.p.s., 7 in. spools. High impedance microphone. High and low ($4\frac{1}{2}$ watts) impedance outputs. Three speakers.

TK919. Similar to TK820 but with refinements, e.g. automatic track reversal.

Accessories

Microphones:	TK9 and TK819, GDM5, GRM1, GRM2. All other models, GCM2, GRM1Z, GXM1.
Telephone Adaptor:	TA1 all models.
Stethoscope Earphones:	STET 1 all models.
Remote Control:	TK12 and TK820, RCF44 and RCF4. All other models RCF22 and RCF2.
Plugs:	All models are designed for Grundig Jack Plugs, J1 and J11.
Spare Leads:	All models are designed for Grundig Spare Leads, SL1 and SL2.

Mains Voltage Adjustment

Pull out the Voltage Selector Plug, C, above, and set so that the figure in the window is nearest your supply voltage:

When 110 shows in window, voltages covered are 105 to 115.

When 200 shows in window, voltages covered are 190 to 210.

When 220 shows in window, voltages covered are 210 to 230.

When 240 shows in window, voltages covered are 230 to 250.

1 2 3 4 5 6 7 8 9 10 11

Distribution Panel Key

1 Mains Connector.
2 Earth Socket on TK12; farther along panel on other models.
3 Mains Voltage Selector.
4 Fuse Holder 105-120 v. mains 2 amp.
5 Fuse Holder 190-250 v. mains 1 amp.
6 Remote Control Socket.
7 Low Impedance Output (2·5 watts for extension speaker or Grundig Stethoscope Headphone, type STET 1 used in conjunction with Grundig

Jack Plug, type J11 or Spare Lead SL2 pushed right in on TK12 separate socket on other models.
8 High Impedance Output (1·2 v.) or external amplifier with above plug or lead assembly pushed half way into socket on TK12; separate socket on other models.
9 Input for Radio L.S./Gram P.U.
10 Input for Diode or TA.
11 Input for Microphone.

Grundig Models TR3, TK5, TK7, TK8

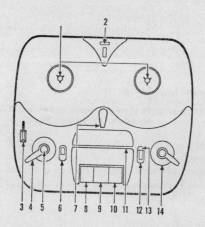

Deck Key

1 Spool Holders (TR3 and TK5 5¾ in., TK7 and TK8, 7 in.).
2 Tape Position Indicator.
3 Speed Control (3¾ and 7½ i.p.s.).
4 Signal Level Control (Recording)/ Volume Control (Play Back).
5 Speaker Control and Cut-out (Recording)/Tone Control (Play Back).
6 Magic Eye.
7 Fast Wind Selector (Left: Wind back, Right: Wind On).

8 Microphone (High Impedance input selector).
9 Diode input selector.
10 Radio L.S./Gram. P.U. input selector
11 Sound Channel.
12 Pause Control.
13 Pause Control Hold Key.
14 Selector Switch:
  ○ Motor Off.
  ↔ Prepare to fast wind.
  ◎ (Gold) Play Back.
  ◎ (Red) Record (After lifting to clear safety lock).

Model Differences

TR3 is a simplified version of the TK5. It has 5¾ in. spools and only one speed, 3¾ i.p.s. It has one input and one output, no input selector, output stage or built-in speaker. It is mounted on a wooden frame.

TK5 has 5¾ in. spools, one speaker, 2½ watts output and one speed, 3¾ i.p.s.

TK7 and TK8 have three speakers, an output of 4½ watts and two speeds, 3¾ and 7½ i.p.s.

v

Accessories

Microphones: GCM3, GDM3Z, GDM111, GRM3Z (all models).

Telephone Adaptor: TA3.

Stethoscope Earphones: STET 3.

Remote Control: Not provided for

Plugs: Grundig 3-pin plug, type J3.

Connecting Leads: All Inputs and Low Impedance Output: SL3.
All Inputs and High and Low Impedance Outputs: SL33.
Input from Grundig radio Diode: SL233.

Mains Voltage Adjustment

Pull out selector plug on voltage selector panel (Righthand compartment at back) and re-set to figure corresponding to your supply voltage—i.e.:

When 110 shows in window, voltages covered are 105 to 115.

When 200 shows in window, voltages covered are 190 to 210.

When 220 shows in window, voltages covered are 210 to 230.

When 240 shows in window, voltages covered are 230 to 250.

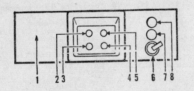

Socket plan

Distribution Panel Key

1 Mains Lead Compartment.
2 Input for Radio L.S./Gram. P.U.
3 Input for Diode/TA3 Adaptor/GMU3 (Pins 1 and 2). High Impedance Output (Pins 3 and 2).
4 Output Socket providing High and Low Impedances in conjunction with Grundig Spare Lead, type SL33. (Black/Yellow wander plugs for Low, 3-7 ohm impedance; Black/Red wander plugs for High Impedance,

e.g. to connect to external amplifier of pick-up sockets of radio or radiogram. Socket also takes Grundig Stethoscope Earphones, STET 3.
5 Input for Microphone. All sockets are designed to take Grundig 3-pin plug, type J3.
6 Mains Voltage Adjustment.
7 Fuse, H.T.
8 Fuses, Mains: 105-115 v., 2 amp., 190-250 v., 1 amp.

Grundig Models TK830/3D, TK16

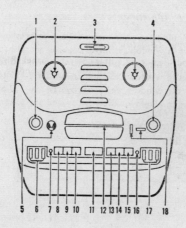

Deck Key

1 Recording Level Control.
2 Spool Holders (TK830/3D, 7 in. spools; TK16, 5¾ in. spools).
3 Speed Control. TK830/3D. 3¾ i.p.s. to left; 7½ i.p.s. to right; TK16, 1⅞ and 3¾ i.p.s. Set before starting.
4 ON/OFF Switch and Volume Control.
5 Magic Eye Recording Level Indicator.
6 Input Selectors: Left, Radio L.S./Gram. P.U.; Centre, Diode; Right, Microphone.
7 Safety/Pause Button.
8 Record.
9 Playback.
10 Track II.
11 Stop Bar.
12 Sound Channel.
13 Track I.
14 Fast Wind ←.
15 Fast Wind →.
16 Loudspeaker and Erase Head Cut-out Button.
17 Tone Controls: Left, Bass; Centre Middle Register; Right, Treble.
18 Tape Position Indicator.

Model Differences

TK16 is a smaller version of the TK830/3D and takes 5¾ in. spools and has two speeds, 1⅞ and 3¾ i.p.s.

Accessories

Microphones: High Impedance; GCM3, GDM3Z, GDM III and GRM3Z.

Telephone Adaptor: TA3.

Stethoscope Earphones: STET 3.

Remote Control: RCF55 (Stop, Start and Back Space).

Plugs: Grundig Type J3, 3-pin plug for input and output connections.

Special: Grundig Distributor Speaker.

Mains Voltage Adjustments

Pull out plug and set to your mains voltage as below:

When 117 shows in window, voltages covered are 110 to 125.

When 150 shows in window, voltages covered are 140 to 160.

When 200 shows in window, voltages covered are 190 to 210.

When 220 shows in window, voltages covered are 210 to 230.

When 240 shows in window, voltages covered are 230 to 250.

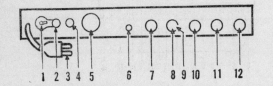

Socket plan

1 2 3 4 5 6 7 8 9 10 11 12

Distribution Panel Key

1   Mains Voltage Adjuster.

2   Fuse, Mains: 190-250 v., 0·8 amp., surge resisting; 140-160 v., 1·0 amp., surge resisting; 110-125 v., 1·25 amp., surge resisting.

3   Mains Lead Connector (Plugs in to lead stored in lid compartment).

4   Fuse, H.T.: 100mA.,

5   The Grundig Distributor Speaker Socket.

6   Earth Socket for Grundig Earth Plug, type J6.

7   Remote Control Socket for Grundig Universal Foot Control, type RCF55.

8   Output I: Pins 1 and 2 (earthed). Low impedance (7½-15 ohm) for extension speaker or Grundig Stethoscope Headphone through Grundig Leads, SL3 or SL33.

9   Output II: Pins 2 (earthed) and 3. High impedance for external amplifier, Grundig Mixer or high impedance headphones (Grundig Lead, SL33).

10   Radio L.S./Gram. P.U. for connexion to Extension Speaker sockets or gramophone pick-up.

11   Diode TA3 for Grundig Telephone Adaptor and Radio set Diode connexion (Pins 1 and 2). High Impedance Output (Pins 3 and 2).

12   Micro. Grundig Condenser Microphone GCM3.

Sockets 8, 9, 10, 11 and 12 are of the Grundig 3-pin type and take Grundig Screened 3-pin plugs, type J3.

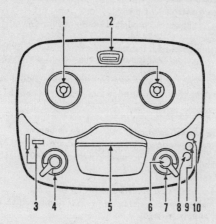

Deck Key

1 Spool Holders (5¾ in. spools).
2 Magic Eye.
3 Tape Position Indicator.
4 Selector Switch:
    ←Fast Wind Right to Left.
    O(Red) Stop motor. (*Always return to this position when not in use.*)
    ▶Play Back or (if any input selector button is held down first) Record.
    ●(Red) Pause.
    → Fast Wind Left to Right.

5 Sound Channel.
6 On-Off Switch and Volume Control (Record)/Tone Control (Play Back). Speaker Cut-out when pulled up.
7 Signal Level Control (Record)/ Volume Control (Playback).
8 Microphone input selector.
9 Radio L.S./Gram. P.U. input selector.
10 Diode input selector.

Model Differences

TK20 has only one speed, 3¾ i.p.s.

TK25 has two speeds, 3¾ and 1⅞ i.p.s. It also has extra controls at the front of the deck for Speed Change, Erase Head Cut-out, and signal fading; the latter also combines the functions of controls 8, 9 and 10 (above) in conjunction with a Record Button which occupies the position of control 9 (above).

TM20 is the chassis version of the TK20. It has no output stage or loudspeaker and cabinet.

TK24 records four tracks at 3¾ i.p.s. and plays back singly or two tracks at once. Facilities for monitoring one track while recording the other.

ix

Accessories

Microphones: GCM3, GDM111, GDM12 (TK24).

Telephone Adaptor: TA3.

Stethoscope Earphones: STET 3.

Remote Control: Not provided for.

Plugs: All models are designed for Grundig 3-pin plug, type J3.

Connecting Leads: All Inputs and Low Impedance Output: SL3.
All Inputs and High and Low Impedance Outputs: SL33.
Input from Grundig radio Diode: SL233.

Monitor Amplifier: MA1 used in conjunction with STET3 for monitoring on 4-track model, TK24.

Mains Voltage Adjustment

1 Loosen grub screw in knob 6 (Deck).

2 Lift off 6 and 7.

3 Pull 4 off spindle.

4 Remove 4 coin-slotted screws and lift off deck top.

5 See that mains adjusting screw is in hole numbered to correspond with your supply voltage.

Position 1 covers voltages from 110 to 125.

Position 2 covers voltages from 190 to 210.

Position 3 covers voltages from 210 to 230.

Position 4 covers voltages from 230 to 250.

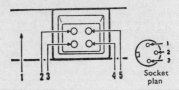

Socket plan

Distribution Panel Key

1 Mains Lead compartment underneath.
2 Input for Radio L.S./Gram. P.U.
3 Input or Diode/TA3/Mixer Unit (Pins 1 and 2). High Impedance Output (Pins 3 and 2).
4 Output Socket providing High and Low Impedance in conjunction with Grundig Spare Lead, type SL33. (Black/Yellow wander plugs for Low, 3-7 ohm impedance; Black/Red e.g. to connect to external amplifier wander plugs for High Impedance— or pick-up sockets of radio or radiogram.)

Socket also takes Grundig Stethoscope Earphone, STET 3.
All sockets are designed to take Grundig 3-pin plug, type J3.
5 Input for High Impedance Microphone.

Under deck:
Fuse, H.T.: 100 mA. surge resisting.
Fuses Mains: 100-125 v., 600 mA.; 180-250 v., 400 mA.
Mains Voltage Adjustment.

Grundig Models TK30, TK35

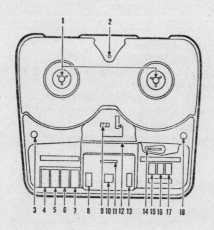

Deck Key

1 Spool Holders (7 in. spools).
2 Pilot Light.
3 Record Button (depress). Loud-speaker and Erase Head Cut-Out Button (turn clockwise and depress
4 Stop Key.
5 Start Key.
6 ◀Fast Wind Right to Left.
7 Fast Wind ▶Left to Right.
8 Volume Control.
9 Tape Position Indicator.
10 Pause Key.

11 Magic Eye Signal Level Indicator.
12 Sound Channel.
13 Signal Level Control (Recording) Tone Control (Playback).
14 Speed Control for $3\frac{3}{4}$ and $7\frac{1}{2}$ i.p.s. (incorporating On-Off Switch on TK35).
15, 16, 17. Input Selectors: Left, Micro-phone; Centre, Diode; Right, Radio L.S./Gram. P.U.
18 On-Off Switch (TK30 only). $1\frac{7}{8}$ i.p.s. Speed Control (TK35 only).

Model Differences

TK30. Two speeds, $3\frac{3}{4}$ and $7\frac{1}{2}$ i.p.s.
TK35. Additional speed of $1\frac{7}{8}$ i.p.s.

Accessories

Microphones: GCM3 and GDM111.

Telephone Adaptor: TA3.

Stethoscope Earphones: STET 3.

Remote Control: RCF30.

Plugs: Grundig Type J3 for all input and output connexions.

Connecting Leads: All Inputs and Low Impedance Outputs: SL3.
All Inputs and High and Low Impedance Out-
puts: SL33.
Input from Grundig radio Diode: SL233.

Mains Voltage Adjustment

Pull out Voltage Selector Plug and re-set so that the figure in the
window is nearest your supply voltage:

When 110 shows in window, voltages covered are 105 to 115.

When 200 shows in window, voltages covered are 190 to 210.

When 220 shows in window, voltages covered are 210 to 230.

When 240 shows in window, voltages covered are 230 to 250.

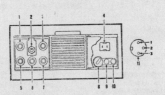

Distribution Panel Key

1 Extension Speaker Output (Pins 1
and 2). High Impedance Output
(Pins 3 and 2).

2 Internal Loudspeaker Switch.

3 Remote Control Socket.

4 Mains Connexion Socket.

5 Microphone Input (High Impedance).

6 Diode Input (Pins 1 and 2). High
Impedance Output (Pins 3 and 2).

7 Radio L.S./Gram. P.U. Input.

8 Mains Voltage Adjustment.

9 Fuse, Mains: 100-125 v., 800 mA.
(TK30), 1A (TK35). 190-250 v.
600 mA. (TK30 and TK35).

10 Fuse H.T.: 125 mA. (TK30), 100 mA.
(TK35).

11 3-pin socket connexions (looking on
face of socket). Pin 2 must always be
soldered to screen of connecting lead.

Grundig Model TK55

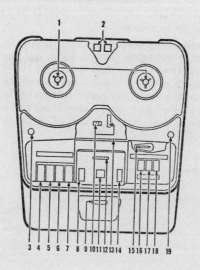

Deck Key

1 Spool Holders (7 in. spools).
2 Mono/Stereo Indicator Light.
3 Record Button (depress)/Loudspeaker and Erase Head Cut-Out Button (turn clockwise and depress).
4 Stop Key.
5 Start Key.
6 ◀ Fast Wind Right to Left.
7 Fast Wind ▶ Left to Right.
8 Volume Control.
9 Mono/Stereo Switch.
10 Tape Position Indicator.

11 Pause Key.
12 Magic Eye Signal Level Indicator.
13 Sound Channel.
14 Signal Level Control (Recording). Tone Control (Playback).
15 Speed Control for $3\frac{3}{4}$ and $7\frac{1}{2}$ i.p.s. incorporating On-Off Switch.
16, 17, 18. Input Selectors: Left, Microphone; Centre, Diode; Right, Radio L.S./Gram. P.U.
19 Press Button for $1\frac{7}{8}$ i.p.s.

xiii

Accessories

Microphones: GCM3 and GDMIII.
Telephone Adaptor: TA3.
Stethoscope Earphones: STET 3.
Remote Control: RCF30.
Plugs: Grundig Type J3 for all input and output connexions.
Connecting Leads: All Inputs and Low Impedance Outputs: SL3.
All Inputs and High and Low Impedance Outputs: SL.33.
Input from Grundig radio Diode: SL233.
Output for Stereo Channel 2 from High Impedance sockets: SL3X.

Mains Voltage Adjustment

Pull out Voltage Selector Plug and re-set so that the figure in the window is nearest your supply voltage:

When 117 shows in window, voltages covered are 110 to 125.
When 150 shows in window, voltages covered are 140 to 160.
When 200 shows in window, voltages covered are 190 to 210.
When 220 shows in window, voltages covered are 210 to 230.
When 240 shows in window, voltages covered are 230 to 250.

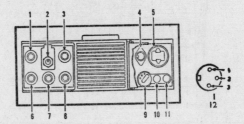

Distribution Panel Key

1 Extension Speaker Output (Pins 1 and 2). High Impedance Output, Mono or L.H. Channel, Stereo (Pins 3 and 2).
2 Internal Loudspeaker Switch.
3 High Impedance Output, Stereo: R.H. Channel, Pins 1 and 2; L.H. Channel, Pins 3 and 2.
4 Remote Control Socket.
5 Mains Connexion Socket.
6 Microphone Input (High Impedance). Pins 1 and 2.

7 Diode Input (Pins 1 and 2), High Impedance Output, Mono (Pins 3 and 2); High Impedance Output, Stereo (Pins 3 and 2).
8 Radio L.S./Gram. P.U. Input.
9 Mains Voltage Adjustment.
10 Fuse, Mains: 110-125 v., 1 amp.; 190-250 v., 500 mA.
11 Fuse, H.T.: 125 mA.
12 3-pin socket connexions (looking on face of socket). Pin 2 must always be soldered to screen of connecting lead.

xiv

Grundig Models TK60, TM60

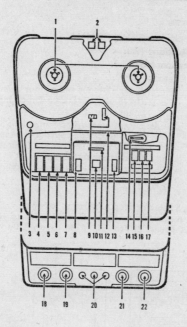

Deck Key

1 Spool Holders (7 in. spools).
2 Mono/Stereo Pilot Light.
3 Recording Button (depress). Loud
   Speaker and Erase Head Cutout
   Button (turn clockwise and depress).
4 Stop Key.
5 Start Key.
6 ◀Fastwind right to left.
7 ▶Fastwind left to right.
8 Mono/Stereo Switch.

9 Tape Position Indicator.
10 Pause Key.
11 Magic Eye Recording Level Indicator
12 Sound Channel.
13 Recording Level Control.
14 Mains On-Off Switch/Speed Control.
15, 16, 17. Input Selectors: Left, Microphone; Centre, Diode; Right, Radio
   L.S./Gram P.U.

Front Panel Key

18 Volume Control.

19 Stereo Balance Control.

20 Microphone Input Sockets: ♀I Mono
   and Left Stereo Channel: ♀II Right

Stereo Channel; I∞I Special Stereo
Microphone, Right and Left Stereo
Channels.
21 Bass Control.
22 Treble Control.

Model Differences

TM60 is a tape deck designed for use in conjunction with the SO131 Radiogram. It can also be used with other stereo reproducers providing the same facilities for amplifying the two high impedance outputs. The input and output connexions are not the same as for the TK60 and there are special installation instructions.

Accessories

Microphones: GCM3, GDM111.

Telephone Adaptor: TA3.

Stethoscope Earphones: STET 3.

Remote Control: RCF30.

Plugs: J3 for all input and output connexions.
J14 (5-pin) for Stereo Diode Input and output connexions.

Connecting Leads: Extension speaker connexion: SL3.
All input and output connexions: SL33/S (Red, Pin 1, Yellow, Pin 3, Black, Pin 2.)
Diode output: SL233.
Special lead for TM60, all connexions: SL233S.

Mains Voltage Adjustment

Adjust Voltage Selector Plug position as follows:
117 in window covers 110 to 125v; 150 covers 140 to 160v; 200 covers 190 to 210v; 220 covers 210 to 230 v; 240 covers 230 to 250 v.

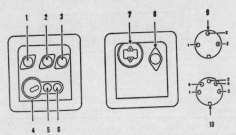

Distribution Panel Key

1   Output Socket. Pins: Mono and Left Stereo, 3 and 2. Right Stereo, 1 and 2.
2   Diode Input. Pins: Mono and Left Stereo, 1 and 2. Right Stereo, 4 and 2. Diode Output. Pins: Mono and Left Stereo, 3 and 2. Right Stereo, 5 and 2.
3   Gram. Pins: Mono and Left Stereo, 3 and 2. Right Stereo, 1 and 2.
4   Mains Voltage Adjustment.
5   Fuse, Mains: 100-125v, 1·25 amp. 190-250v, 800 mA.
6   Fuse, HT: 250 mA.
7   Mains Connecting Socket.
8   Remote Control.
9   3-pin socket connexions (looking on face of socket).
10 5-pin Diode socket connexions (looking on face of socket) (Note, this socket will take a standard 3-pin J3 plug.)

Grundig Model TK14, TK14L, TK18, TK18L

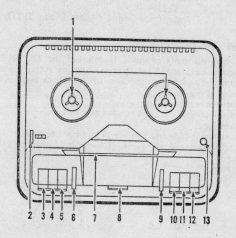

Deck Key

1 Spool Holders (5¾ in. spools).
2 Tape Position Indicator.
3 ◀ Fast Wind Right to Left.
4 Pause Key.
5 Input Selector (depress to record from Microphone; press again to restore and record from Radio or Gram).
6 On-Off Switch/Tone Control.
7 Sound Channel.

8 Magic Eye Signal Level Indicator, TK14; Pilot light, TK18.
9 Recording Level (Record)/Volume (Playback), TK14; Volume (Playback), TK18.
10 Start Key.
11 Stop Key.
12 ▶ Fast Wind Left to Right.
13 Record Button.

Model Differences

TK18 and the De Luxe model TK18L are equipped with a 'magic ear' automatic sensing unit for automatically setting the recording level. They have no recording level control or magic eye.

Accessories

Microphone: GDM18.
Telephone Adaptor: TA3.
Earphones: STET 3 stethoscope earphones.
SE3 single earpiece with earclip.
STET stethoscope attachment for above.

Plugs: Grundig Type J3 for all input and output connexions.

Grundig Type J21 for extension speaker lead (TK18).

Connecting Leads: All inputs and Low Impedance Outputs, SL3.

All inputs and Low and High Impedance Outputs, SL33.

Input from Grundig Radio Diode, SL233.

Mains Voltage Adjustments

The TK14 as normally supplied is adjusted for 220 to 240v A.C. This setting can be changed to cover A.C. mains voltages from 200 to 220 by moving the position of the mains fuse (visible through the window in the bottom of the case).

1 Disconnect recorder from mains.
2 Remove 4 screws through rubber feet in base.
3 Remove base.
4 Pull out mains fuse and replace in appropriate holder. The values are clearly printed alongside. (Do not confuse with HT fuse on right.)
5 Refit base.

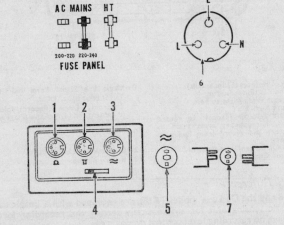

Distribution Panel Key

1 Microphone Input.
2 Radio and Gram P.U. Input (Pins 1 and 2); High Impedance Output (Pins 3 and 2).
3 Extension Speaker Output (Pins 1 and 2); High Impedance Output (Pins 3 and 2) (TK14).
4 Built-in Speaker Switch (TK14).
5 Extension Speaker output, replacing 3 and 4 above on TK18.
6 Mains Connector Socket. Neutral to N; Live to L; Earth to E (TK14).
7 Method of inserting J21 plug into TK18 extension speaker output socket (5) to reproduce through both internal and extension speakers (left) or through extension speaker only (right).

Working Instructions—*in a nutshell*

CONNECT TO MAINS with lead plugged in to separate 3-pin socket at back of case (TK14) or connected under lid in compartment in base (TK18).

LOAD WITH TAPE—full spool on left, empty spool on right.

CONNECT INPUT by inserting 3-pin plug on end of appropriate lead into socket on distribution panel at back of case. (Microphone socket marked $\underline{O}$ Radio/Gram P.U. socket marked $\mathbb{T}$).

SWITCH ON by turning On-Off Switch/Tone Control thumbwheel on left of Magic Eye away from you until it clicks. Wait for Magic Eye to light up (TK14) or green pilot light (TK18).

RECORD (a) From Microphone: depress Input Selector Key (marked MICRO). From Radio or Gram P.U.: leave key *up*. If already *down*, restore by pressing once. (b) Press and hold down Record Button on right of deck. (c) Press and hold Start Key. (d) Release Record Button and then release Start Key to start tape. (e) Press Pause Key (marked Temp. Stop). (f) Adjust Recording Level/Volume Control thumbwheel on right of Magic Eye (marked Volume) until Magic Eye just closes on strongest signals. (g) Release Pause Key to re-start tape. (h) Press Stop Key to end recording. (If you only want to pause, press Pause Key.)

Notes. (b) On the TK18 the Record Button must be turned clockwise before depressing.
(f) On the TK18 the recording level is set automatically and this step is not required.

WIND BACK (a) Press Fastwind Key ◀. (b) When all tape has been wound back on to left spool, press Stop Key.

PLAYBACK (a) Press Start Key. (b) Adjust volume with Recording Level/Volume Control. (c) Adjust tone with On-Off Switch/Tone Control thumbwheel on left of Magic Eye. (d) Press Stop Key to finish playback. (If you only want to pause, press Pause Key.)

SWITCH OFF (a) Check that you have pressed Stop Key. (b) Turn On-Off Switch/Tone Control thumbwheel towards you until it clicks.

The Grundig Cub, TK1

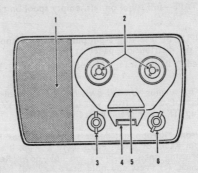

Deck Key

1 Loud Speaker.
2 Spool Holders (3-in. spools).
3 Selector Switch: ● (red) Off.
    ▼ (black) Rewind.
    ○ (green) Play Back.
    ○ (black) Record.

4 Pause Bar.
5 Sound Channel.
6 Signal Level Control (Record)/volume Control (Play Back).

Model Differences

TK1 has a magic eye recording level indicator, capstan drive, $3\frac{3}{4}$ i.p.s. governed tape speed, record/safety lock, and controls for monitor speaker (record) and tone (playback) incorporated in right hand control. Different selector switch colour code.

Accessories

Microphone: Grundig Type GM1 (Cub), GM1 (TK1).
    Tape: Grundig Type TDP 6.
    Plugs: For external battery, Type J12 (Cub), Type J17 (TK1).
    For connexion to Diode socket of Grundig Radio or Tape Recorder: J3.
Connecting Lead (Cub): All Input and Output Connexions Type SL154.
    Special recording lead SL142R.
    (TK1): All Input and Output Connexion—S, Type SL144.
    Special recording lead SL132R.
Batteries: 1·5v cells: Ever Ready Type LPU2 or equiv.
    (Cub) 3v battery: Ever Ready Type 8 or equivalent.
    (TK1) 4x1·5-volt Ray-o-vac Industrial Type 3LP.
    2x1·5-volt Ray-o-vac Type 1LP.
Mains Pack CMP1: For running Cub off main electricity supply.

Distribution Panel Key

Cub

1 Socket for external 6 v. battery connexion.
2 Microphone/Radio/Gram P.U. Input (Pins 1 and 2). High Impedance Output (Pins 4 and 3).

TKI

1 Microphone/Radio/Gram P.U. Input (Pins 1 & 2). High Impedance Output (Pins 3 & 2).
2 Socket for external 6 v battery connexion, 1 to battery Positive 2 to battery Negative.

Battery Arrangement

View looking on back with baseplate removed.
Brass caps on batteries to be in positions shown thus ⊨

Cub

TKI

Working Instructions—in a nutshell

REMOVE LID and lay tape recorder flat on level surface.

LOAD TAPE (3-in. spool of TDP6)—full spool on left, empty spool on right. Wind until all leader tape has passed through sound channel.

CONNECT INPUT—plug microphone or radio/gram recording lead into front socket on left side of Cub, right side of TKI.

RECORD (a) Turn Recording Level/Volume Control knob (right) to 2 (Position may be adjusted after trial). (b) Turn Selector Switch to Record position on extreme right. (c) Return Selector Switch to Stop position to finish recording. (If you only want to pause, press Stop Bar in front of Sound Channel.)

WIND BACK (a) Turn Selector Switch to Rewind position on extreme left. (b) Turn Switch back to Stop position when all tape is back on left hand spool.

PLAYBACK (a) Turn Selector Switch to Playback position. (b) Adjust volume with Recording Level/Volume control knob. (c) Return to Stop position at end of record. (If you only want to pause, press Stop Bar.)

SWITCH OFF—Make sure that Selector Switch is in Stop position.

Grundig Model TK40

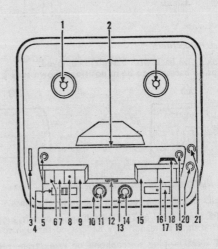

Deck Key

1 Spool Holders (5 in. spools, 7 in. with cover off).
2 Sound Channel.
3 Speed Selector 1⅞, 3¾, 7½. ON-OFF Switch.
4 Fastwind.
5 Tape Cleaner Release Button.
6 Record.
7 Trick (Superimpose).
8 Stop.
9 Start.
10 Monitor Speaker Volume Control (all inputs).
11 Recording Level Control (radio L.S. and pickup inputs).
Tone Control (playback).

Monitor Speaker ON-OFF Switch (pull up for OFF).
12 Magic Eye Recording Level Indicator.
13 Input Selector Arm for Microphone (arm to left) and Radio Diode (arm to right).
14 Recording Level Control Knob for Microphone and Radio Diode Inputs. Volume Control (playback).
15 Selector for Tracks 1-2.
16 Selector for Tracks 3-4.
17 Pause Control.
18 Tape Position Indicator.
19 Tape Position Indicator Reset.
20 Cine Sound Head Input.
21 Microphone Input.

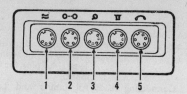

Rear Panel Sockets

1 High Impedance Output for Monitor Amplifier, MA2 (Pins 4, 3, 5). Low Impedance Output for 3-5 Ω Extension L.S. (Pins 2 and 1).
2 Socket for Remote Control, RCF30 (Pins 2 and 1).
3 Input from Radio Extension Speaker Terminals or Pickup (Pins 2 and 1).
4 Diode Connexion (High Impedance Output Pins 3 and 2; Diode Input. Pins 2 and 1).
5 Telephone Coupling Unit Socket (Pins 5, 4, 3, 1).

Mains Connector Socket (*on left side of case*)

Takes 3-pin power plug on end of mains connecting lead. L = Live (Red wire); E = Earth (Green wire); N = Neutral (Black wire).

Microphone Socket (*on right of deck, marked*)

For GDM18 Microphone: Pin 1, Microphone, Pin 2, Screen.

Cine Socket (*on right of deck, marked* ꝅ)

For J14 plug connected to lead from cine projector sound head: Pin 1, Record, playback head; Pin 2; Pin 3, Erase Head; Pins 4 and 5 Screen.

Accessories

Microphone: GDM18.
Telephone Coupling Unit: In preparation.

xxiii

Earphone: SE3.

Stethoscope Attachment for above: STET.

Remote Control Solenoid: F40.

Foot Control RCF30 for use in conjunction with Solenoid F40.

Plugs: Type J3 for all 3-pin connexions.
Type J14, 5 pins, for Cine Tape Head Input.

Connecting Leads: All inputs and Low Impedance outputs, SL3.
All inputs and High and Low Impedance outputs SL33.
Connexions to Grundig Radio Diode socket SL233.

Monitor Amplifier: MA2 for synchronising recordings on separate tracks.

Mains Voltage Adjustment

1 Take out four screws in rubber feet of case and remove cover.
2 Slacken screws holding links in mains adjuster.
3 Re-arrange links according to code below to suit your supply voltage.
4 Tighten screws and replace cover.

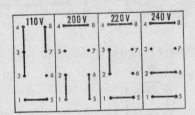

Working Instructions—in a nutshell

CONNECT TO MAINS with mains lead (stowed in compartment in base) plugged in to special 3-pin socket in left hand side of case.

LOAD WITH TAPE—full spool on left, empty spool on right. Wind on to right spool until metallised stop foil is through sound channel.

CONNECT INPUT from microphone, pickup, radio, etc., by inserting 3-pin plug (or special telephone coupling plug if recording from telephone) into appropriate marked socket on top of deck (microphone) or rear panel (other sources).

SELECT INPUT—For microphone and radio diode use arm on right hand control. Pickup or radio extension speaker input is automatically selected when plug is inserted in socket.

SELECT SPEED—By turning speed selector wheel on left of deck to $- 1\frac{7}{8}$, $= 3\frac{3}{4}$ or $\equiv 7\frac{1}{2}$.

SWITCH ON—The recorder switches on automatically when you set the speed selector.

SELECT TRACK by pressing the appropriate Track Selector Key (12- or 3-4) on right of deck.

RECORD (a) Pull up left hand control knob to switch off monitor speaker if recording from microphone. (b) Press Record Key (left keybank). (c) Adjust recording level with right hand knob (microphone and radio diode) or left hand knob (pickup or radio L.S.). (d) Press Start Key (left keybank). (e) To end recording, press Stop Key (left keybank). (For short pause press Pause Key on right.)

WIND BACK (a) Move Fastwind slider (left of deck) to left. (b) When all tape is back on left hand spool, press Stop Key.

PLAY BACK (a) Push down knob on left control to switch on speaker. (b) Press Start Key. (c) Adjust volume with knob on right hand control. (d) Adjust tone with knob on left hand control. (e) Press Stop Key to end playback. For short pause, press Pause Key on right.

SWITCH OFF (a) Press Stop Key to return all other controls to normal. (b) Turn speed selector—On-Off Switch to intermediate O between any two speed markings.

Radio Connexions

TO RECORD FROM RADIO L.S. TERMINALS (a) Push 3-pin plug on end of screened lead SL3 supplied into socket marked Q on rear panel. (b) Connect red and black leads to output sockets of radio receiver with suitable plugs, red to live, black to screen or chassis. Reverse black and red leads if hum interferes. (c) Record as above using left hand knob for adjusting recording level, and control arm for controlling volume through speaker or knob for switching it off.

TO RECORD FROM RADIO DIODE (a) Push 3-pin plug on end of screened lead SL33 (from your dealer) into diode socket in rear panel, marked ∀ (b) Connect red and black leads to radio diode output, red to live and black to screen or chassis. (c) Turn right hand control knob to diode position at right, marked ∀ (d) Record as above, using right hand control knob for recording level and left hand control arm for speaker or pull up knob to switch off speaker.

TO PLAY BACK THROUGH RADIO (a) Push 3-pin plug on end of screened lead SL33 into diode socket on rear panel. (b) Connect black and yellow leads to Pickup input sockets on radio. (c) Switch radio to GRAM. (d) Pull up speaker cut-out knob (left hand control) on TK40 and playback tape as above, with volume control knob turned about one-third way and using volume and tone controls on radio.

Gramophone Connexions

TO RECORD FROM GRAMOPHONE PICKUP (a) Push 3-pin plug on end of screened lead SL3 into pickup socket (marked Ω on rear panel). (b) Connect screened lead to pickup terminals, red to live, black to screen or chassis. (c) Start turntable and lower stylus on to disc. (d) Record as for radio L.S. above.

Extension Speaker Connexions

TO PLAY BACK THROUGH AN EXTENSION SPEAKER (3-5 OHM)
(a) Push 3-pin plug on end of screened connecting lead SL3 into output socket marked ▬ on rear panel. (b) Connect red and black plugs to extension speaker terminals. (c) Pull up speaker cut-out knob (left hand control). (d) Play back as above.

Connexions to Grundig Radio

Most Grundig radio sets and radiograms (and some other makes) are fitted with a special diode socket for connecting to a tape recorder. To connect the TK-40 to this socket you need the special screened lead SL233 which has a 3-pin plug on each end. With this lead plugged into the diode sockets of the radio and the TK40 you can record a programme from, and play it back through the radio set by using the normal switching arrangements.

Synchronising

With the TK40 four-track recorder you can listen to a record previously made on one track while you make a second record to synchronise with it on the associated track. For this you need the Monitor Amplifier MA2 connected to the Stet 3 Earphone.
(a) Make first record on Track 1 or 2 and wind back to start. (b) Connect MA2 to output socket ▬ on rear panel with plug on end of lead. (c) Press Key for second pair of tracks. (d) Record second item, normally. While recording you will hear the first item in the earphone, enabling you to synchronise the two recordings.

Playing Back Two Tracks at Once

When you have made two synchronised recordings as above, you can play them both back together by pressing both track keys at the same time. You hear both records through the built-in speaker and can control them as for a normal single track record.

Cine Socket

This socket can be connected (by J14 plug) to the Record/Play and Erase heads of a cine projector equipped for magnetic tape recording and playback.
(a) Plug connecting lead (short and well screened) into cine projector and TK40. (b) Return both track selector keys to neutral by pressing and releasing both at once. (c) Record and play back using normal controls on recorder and projector.

Grundig Model TK41

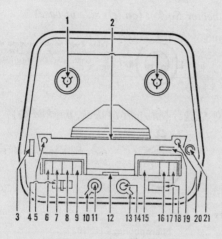

Deck Key

1	Spool Holders (5 in. spools, 7 in. with cover off).	11	Treble Tone Control Knob.
2	Sound Channel.	12	Magic Eye Recording Level Indicator.
3	ON-OFF Switch/Speed Selector 1⅞, 3¾, 7½ i.p.s.	13	Speaker ON-OFF Switch/Arm.
4	Fastwind.	14	Recording Level Control Knob (record)/Volume Control (playback).
5	Tape Cleaner Release Button.	15	Straight Through Amplifier.
6	Record.	16	Microphone Input Selector.
7	Trick (Superimpose).	17	Diode and Radio/Pickup Selector.
8	Stop.	18	Pause Key.
9	Start.	19	Tape Position Indicator.
10	Bass Tone Control Arm/Monitor Speaker Volume.	20	Tape Position Indicator Reset
		21	Microphone Socket.

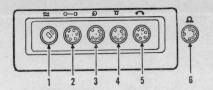

Rear Panel Sockets

1 Low Impedance Output for 3-7 ohm Extension L.S.
2 Socket for Remote Control, RCF30 (Pins 2 and 1).
3 Input from Radio Extension Speaker Terminals or Pickup (Pins 1 and 2).

4 Diode Connexion (High Impedance Output, Pins 3 and 2; Diode input, Pins 2 and 1).
5 Telephone Coupling Unit Socket (Pins 5, 4, 3, 1).

Mains Connector Socket (*on left side of case*)

Takes 3-pin power plug on end o mains connecting lead. L = Live (Red wire); E = Earth (Green wire); N = Neutral (Black wire).

Microphone Socket (*on right of deck, marked* ◯)

For GDM18 Microphone: Pin 1, Microphone, Pin 2, Screen.

Accessories

Microphone: GDM18.
Telephone Coupling Unit: In preparation.
Earphones: SE3.
Stethoscope Attachment for above: STET
Remote Control Solenoid: F40
Foot Control RCF30 for use in conjunction with Solenoid F40.
Plugs: Type J3 for all 3-pin connexions.
Type J21 for extension speaker connexions.
Connecting Leads: All inputs and Low Impedance outputs, SL3.
All inputs and High and Low Impedance outputs SL33.
Connexions to Grundig Radio Diode socket SL233.

Mains Voltage Adjustment

1 Take out four screws in rubber feet of case and remove cover.
2 Slacken screws holding links in mains adjuster.
3 Re-arrange links according to code below to suit your supply voltage.
4 Tighten screws and replace cover.

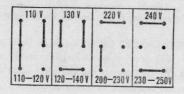

Working Instructions—in a nutshell

CONNECT TO MAINS with mains lead (stowed in compartment in base) plugged in to special 3-pin socket in left hand side of case.

LOAD WITH TAPE—full spool on left, empty spool on right. Wind on to right spool until metallised stop foil is through sound channel.

CONNECT INPUT from microphone, pickup, radio, etc., by inserting 3-pin plug (or special telephone coupling plug if recording from telephone) into appropriate marked socket on top of deck (microphone) or rear panel (other sources).

SELECT INPUT—By depressing appropriate MICRO or RADIO input selector key on right of deck.

SELECT SPEED—By turning speed selector wheel on left of deck to - $1\frac{7}{8}$, —$3\frac{3}{4}$ or ≡ $7\frac{1}{2}$.

SWITCH ON—The recorder switches on automatically when you set the speed selector.

RECORD (a) Switch off speaker with right hand control arm if recording from microphone. (b) Press RECORD Key, (left key bank). (c) Adjust recording level with right hand control knob. (d) Press START Key (left key bank). (e) To end recording, press STOP Key (left key bank). For short pause, press PAUSE Key on right.

WIND BACK (a) Move Fastwind slider (left of deck) to left. (b) When all tape is back on left hand spool, press STOP Key.

PLAY BACK (a) Switch on speaker with right hand control arm. (b) Press START Key. (c) Adjust volume with knob on right hand control. (d) Adjust tone with left hand control (arm, treble; knob, bass). (e) Press STOP key to end playback. For short pause, press PAUSE key.

SWITCH OFF (a) Press STOP key to return all other controls to normal. (b) Turn speed selector ON-OFF switch to intermediate 0 between any two speed markings.

Radio Connections

TO RECORD FROM RADIO L.S. TERMINALS (a) Push 3-pin plug on end of screened lead SL3 supplied into socket marked ♋ on rear panel. (b) Connect red and black leads to output sockets of radio receiver with suitable plugs, red to live, black to screen or chassis. (c) Press RADIO Key on extreme right. (d) Record as above, using right hand control knob for adjusting recording level and left hand control arm for adjusting monitor speaker volume—or right hand control arm for switching it off. Reverse black and red leads if hum interferes.

TO RECORD FROM RADIO DIODE (a) Push 3-pin plug on end of screened lead SL33 (from your dealer) into diode socket on rear panel, marked 🍸 (b) Connect red and black leads to radio diode output, red to live and black to screen or chassis. (c) Press RADIO Key. (d) Record as above.

TO PLAY BACK THROUGH RADIO (a) Push 3-pin plug on end of screened lead SL33 into diode socket on rear panel. (b) Connect black and yellow leads to Pickup input sockets on radio. (c) Switch radio to GRAM. (d) Switch on speaker with arm on right hand control. (e) Set TK41 volume control knob mid-way and play back using volume control on radio.

Gramophone Connexions

TO RECORD FROM GRAMOPHONE PICKUP (a) Push 3-pin plug on end of screened lead SL3 into pickup socket marked ♋ on rear panel. (b) Connect screened lead to pick-up terminals, red to live, black to screen or chassis. (c) Press RADIO Key on right. (d) Start turntable and lower stylus on to disc. (e) Record as from Radio L.S. above.

Extension Speaker Connexions

TO PLAY BACK THROUGH EXTENSION SPEAKER (3-7 OHM) (a) Connect extension speaker by twin lead to TK41 by 2-pin plug Type J21 pushed into socket marked ▄▄▄ on rear panel. (b) Switch off built-in speaker with arm on right hand control. (c) Play back as above.

Note: The full output power of 7 watts is available when connected to an extension speaker. The output is restricted when playing back through the built-in speaker to prevent it from being damaged by overloading.

Connexions to Grundig Radio

Most Grundig radio sets and radiograms (and some other makes) are fitted with a special diode socket for connecting to a tape recorder. To connect the TK41 to this socket you need the special screened lead SL233 which has a 3-pin plug on each end. With this lead plugged into the diode sockets of the radio and the TK41 you can record a programme from, and play it back through the radio set by using the normal switching arrangements.

XXX

Grundig Model TK46, TM45

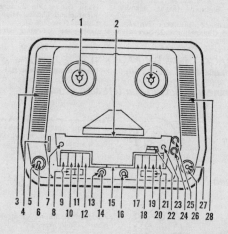

Deck Key

1 Spool Holders (7 in. spools).
2 Sound Channel.
3 Left Channel Speaker.
4 Speed Selector 1⅞, 3¾, 7½ i.p.s./ ON-OFF Switch.
5 Feed back (Black Line).
6 Recording Level (Red Line) (Pull to up give individual control of each channel).
7 Tape Cleaner Release Button.
8 Fastwind.
9 Record Key, Tracks 1-2.
10 Record Key, Tracks 3-4.
11 Stop.
12 Start.
13 Bass Control Arm.
14 Treble Control Knob.
15 Recording Level Indicator.
16 Input Selector Arm: Microphone, left; Telephone Adaptor, centre; Radio Diode, right.

17 Tape Monitor Key (Marked CON.)
18 Synchronous Monitor Key. (Marked SYN.)
19 Track 1-2 playback Key.
20 Track 3-4 playback Key.
21 Pause Key.
22 Tape Position Indicator.
23 Tape Position Indicator Reset.
24 Microphone Input for Right Hand Stereo Channel. (Pin 1, Microphone; Pin 2, Screen.)
25 Microphone Input for Mono recordings and for Left Hand Stereo Channel (Pin 1, Microphone; Pin 2 Screen.)
26 Volume Control (Black line) for Left Hand Speaker.
27 Volume Control (Red line) for Right Hand Speaker.
28 Right channel Speaker.

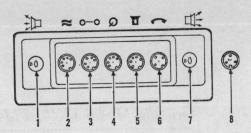

Rear Panel Key

1 Extension Speaker (Right Hand Speaker).
2 High Impedance output. (Pin 1, Right Hand Channel; Pin 2, Screen; Pin 3, Left Hand Channel).
3 Remote Control. (Pins 2 and 1).
4 Input for Pickup. (Pin 1, Right Hand; Pin 2, Screen; Pin 3, Left Hand.)
5 Input for Radio Diode. (Pin 1, Left Hand Diode; Pin 2, Screen; Pin 4, Right Hand Diode.) High Impedance Output (Pins 3 & 5).
6 Telephone Coupling Unit. (Pin 1, Unit; Pin 2, Screen.)
7 Extension Speaker (Left Hand Speaker).
8 Mains Connector. (Live, left pin; Earth, top pin; Neutral, right pin).

Mains Connector Socket *(on left side of case)*

Takes 3-pin power plug on end of mains connecting lead. L = Live (Red wire); E = Earth (Green wire); N = Neutral (Black wire).

Microphone Sockets *(on right of deck, marked* Ω*)*

For GDM18 Microphones: Pin 1, Microphone, Pin 2, Screen.

Model Differences
TM45 is the deck only of TK46.

Accessories

Microphones: Mono, GDM12, GDM18 on either channel; Stereo, GDM12 or GDM 18's on each channel or special Grundig Stereo microphone, GDSM 202.

Telephone Coupling Unit: In preparation.

Earphone: SE3.

Stethoscope Attachment for above: STET.

Remote Control Solenoid: F40

Foot Control RCF30 for use in conjunction with Solenoid F40.

Plugs: Type J3 for all 3-pin connections.
Type J14 for all 3- or 5-pin connexions.
Type J21 for all 2-pin connexions to extension speakers or earphones.

Connecting Leads: Type SL155 for all TK46 input and high impedance output connexions.
Type SL233 for connexions to Grundig Radio fitted with diode connexion socket.
Type SL255 for connexions to Grundig Stereo equipment fitted with 5-pin diode connexion socket.
Type SL33-S for all input or high impedance output connexions.

Mains Voltage Adjustment

1 Take out four screws in rubber feet of case and remove cover.
2 Slacken screws holding links in mains adjuster.
3 Re-arrange links according to code below to suit your supply voltage.
4 Tighten screws and replace cover.

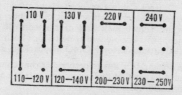

Working Instructions—in a nutshell

CONNECT TO MAINS with separate mains lead plugged in to special 3-pin socket at side of rear panel.

LOAD WITH TAPE—full spool on left, empty spool on right. Wind on to right spool until metallised stop foil is through sound channel.

CONNECT INPUT from microphone, pickup, radio, etc., by inserting 3-pin plug (or special telephone coupling plug if recording from telephone) into appropriate marked socket on top of deck (microphone) or rear panel (other sources).

SELECT INPUT with control switch on right: Microphone, arm to left; Telephone Unit, centre; Radio-pick-up, arm to right.
SELECT SPEED—By turning speed selector when on left of deck to
- $1\frac{7}{8}$, = $3\frac{3}{4}$ or $\equiv 7\frac{1}{2}$.

SWITCH ON—The recorder switches on automatically when you set the speed selector.

SELECT TRACK by depressing appropriate track selector key—left keybank, record; right keybank, playback.

RECORD (MONO) (a) Turn monitor speaker control knobs on right fully anticlockwise unless you want to hear the signal as you record it (see Monitoring, below). (b) Press RECORD key (left keybank), numbered for the track you want to use. (c) Adjust recording level with upper control knob on left (red line). (d) Press START key (Left keybank). (e) To end recording press STOP key (left keybank). For short pause press PAUSE key on right.

RECORD (STEREO) (a) Turn monitor speaker control knobs on right fully anticlockwise unless you want to hear the signal as you record it (See Monitoring, below). (b) Press both RECORD Keys (left keybank). (c) Continue as for Mono recording (c) above.

WIND BACK. (a) Move Fastwind slider (left of deck) to left. (b) When all tape is back on left hand spool, press STOP key.

PLAY BACK (MONO) (a) Press playback key 1-2 or 3-4 according to track to be played. (b) Press START key to start tape. (c) Control volume through both speakers with twin control knob on right. (d) Control treble tone with knob on left hand control; bass with arm on same control. (e) Press STOP key to end playback. For short pause, press PAUSE key on left.

PLAYBACK (STEREO) (a) Press both 1-2 and 3-4 playback keys (right keybank). (b) Control volume of both channels with twin control knob on right. (c) If the channels are out of balance, pull up the top control knob and adjust each half of the control independently. When in balance, press knob down again to lock both halves together once more. (d) Continue as for MONO playback (d) above.

SWITCH OFF. (a) Press STOP key to return all other controls to normal. (b) Turn speed selector-ON-OFF switch to intermediate 0 between any two speed markings.

Radio Connexions

TO RECORD FROM RADIO L.S. TERMINALS (a) Push 5-pin plug on end of lead SL155 (supplied with machine) into socket marked ☊ on rear panel. (b) Connect free ends of lead to radio extension speaker sockets with suitable plugs: MONO, Yellow and Screen; STEREO, Yellow and Screen to Left channel output, Mauve and Screen to Right channel output. (c) Turn Input Selector arm (right hand control) to ☊. (d) Record as above. Reverse extension speaker connexions if hum is present on playback.

TO RECORD FROM RADIO DIODE (a) Push 5-pin plug on end of lead SL255 into socket marked ▽ on rear panel. (b) Connect free ends of lead to diode socket on radio (p. 49) with a suitable plug or plugs, wired: MONO, and L.H. channel Stereo, Grey to Input,

Yellow to Output, screen to chassis; STEREO R.H. channel, Pink to input, Mauve to Output, screen to chassis. (c) Turn Input Selector arm to 📺 . (d) Record as above.

TO PLAY BACK THROUGH RADIO OR EXTERNAL AMPLIFIER.

(a) Push 5-pin plug on end of lead SL155 into socket marked ▬▬ on rear panel. (b) Connect free ends of lead to radio pick up sockets or amplifier input, Yellow to pickup, screen to chassis. (c) Switch radio to GRAM. (d) Play back as above, using radio volume and tone controls with TK46 volume controls turned down and tone controls at minimum (for a start).

Gramophone Connexions

TO RECORD FROM GRAMOPHONE PICK UP (a) Push 5-pin plug on end of screened lead SL155 into socket marked on 𝛀 rear panel. (b) Connect free ends of lead to pick up cartridge: MONO and LH Channel Stereo, Yellow and Screen; STEREO RH Channel, Mauve and Screen. (c) Start turntable and lower stylus on to disc. (d) Record as for radio L.S. above.

Extension Speaker (or Earphone) Connexions

TO PLAY BACK THROUGH EXTENSION SPEAKER(S) Connect extension speakers (3-7 Ohm) to TK46 by twin flex connected to J21 plugs (supplied). (b) Push plugs in to extension speaker sockets on back panel; right hand speaker to right hand socket. (Inserting plugs automatically disconnects internal speakers.) (c) Play back as above.

Connexions to Grundig Radio With Diode Socket

Most Grundig radio sets and radiograms—and some other makes—are fitted with a special Diode socket for connecting to a tape recorder via screened lead SL255 (for 5-pin sockets) or SL233 (for 3-pin sockets). When the TK46 is connected in this way you can record a programme *from* and play it back *through* the radio set by using the normal switching arrangements for recording and playing back through the radio diode (above).

Monitoring

DIRECT (a) Record normally. (b) Press playback key corresponding to the recording key in use. (c) Control volume as for playback, above.

VIA TAPE (a) Record normally. (b) Press playback key for 'spare' track, *i.e.* if recording on 1-2, press key 3-4. (c) Press CON key. (d) Control volume as for playback, above. (e) To switch to direct monitoring, press playback key corresponding to recording key in use.

Multiple Recording

SYNCHRONISING TWO RECORDINGS (a) Make first record normally. (b) Play back with appropriate track key and SYN key pressed and volume set so that record is just audible. (c) Press record key for opposite track and record normally, synchronising second recording with sound of first. (d) To hear effect, play back with playback keys for both tracks pressed (giving independent control of each recording through its own speaker) or both keys up, giving common control of both recordings through both speakers.

CROSS RECORDING (a) Make first record normally. (b) Turn recording level control knob on left (red line) fully anticlockwise and pull up. (c) Press 'opposite' recording key and start tape. (d) Adjust recording level to give normal indication on magic eye with lower control knob (black line). (This operation is to find the correct recording level.) (e) Wind back tape to start and repeat operation at correct recording level. (f) The recording will now be transferred to the 'opposite' track. It may be monitored while recording by pressing the appropriate playback key and the new recording may be played back in the usual way.

COMBINING TWO RECORDINGS (a) Proceed as for Cross Recording above. (b) Before making the final cross recording, connect input for second recording with appropriate recording key pressed and adjust recording level for this with the upper recording level control (red line). (c) With both cross and direct recording level controls set, start tape and make second recording normally while monitoring, the first recording through built in speakers or headphones as above. Both recordings will now be combined on the new track.

ADDING FURTHER RECORDINGS (a) Proceed as for Cross Recording and Combining Two Recordings, above, ending with the 1st and 2nd recordings combined on one track. (b) Wind back to beginning. (c) Press recording key for new track and both playback keys. (d) Establish recording level for cross-recording on to new track with lower control knob (black line). (e) Connect new signal source. (f) Establish recording level for new signal source with upper control knob (red line). (g) Set both recording level controls and make final recording run with record key pressed for new track and both playback keys pressed. (h) Monitor old double recording with volume control knob for old track and new triple recording with volume control knob for new track.

This process may be repeated, cross-recording from one track to another and combining with a further recording each time.

ADDING ECHO EFFECT (a) Press record key and adjust recording level. (b) Pull up recording level control. (c) Press playback key for opposite track and CON key. (d) Set cross-recording level control about midway. (e) Make test recording and re-set cross-recording level control as required to give suitable strength of echo.

Note: The echo delay varies with the tape speed: 210, 420 or 840 milliseconds with tape speeds of $7\frac{1}{2}$, $3\frac{3}{4}$ or $1\frac{7}{8}$ i.p.s. respectively.

Grundig Model TK/TS/TM340 TK341, TK/TS320

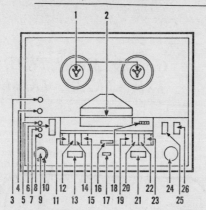

Deck Key

<table>
<tr><td>I</td><td>Spool Holders (7 in. spools).</td></tr>
<tr><td>2</td><td>Sound Channel.</td></tr>
<tr><td>3</td><td>Microphone Input: Mono and Left Channel Stereo.</td></tr>
<tr><td>4</td><td>Microphone Input: Right Channel Stereo.</td></tr>
<tr><td>5</td><td>Straight through Amplifier Switch.</td></tr>
<tr><td>6</td><td>Mains On-Off and speeds 1⅞, 3¾ and 7½.</td></tr>
</table>

7 Input Selector, Microphone.
8 Input Selector, Radio/gram Pickup. — Depress both to record from telephone with output stages switched off.

9 Recording Level, mono and stereo. (Pull up to switch to 10, below.)
10 Recording Level, multiple synchronous effects. (With 9, above, pulled up.)
11 Record Channels 1 and 2.

12 Record Channels 3 and 4.
13 Fast Wind.
14 Stop.
15 Start.
16 Recording Level Indicator.
17 Pilot Lamp.
18 Tape Position Indicator.
19 (CON) Tape monitoring on mono recordings and adding echoes.
20 (SYN) Monitoring synchronous recordings.
21 Pause.
22 Playback, Channels 1 and 2.
23 Playback, Channels 3 and 4.
24 Direct monitoring mono and stereo, recording/Volume and stereo balance, playback. (Pull up top knob to free control for independent adjustment of channels.)
25 Bass.
26 Treble.

The principal features and the method of operation of these stereo recorders are identical with the TK46. However, the layout and arrangement of the controls, shown above, are different on this, the later model.

Model Differences

TK341 and TM340 have 3 separate input selector buttons: Micro, Radio and Gram and no straight through amplifier.

TK340 has two 12 watt output stages.

TK341 is a portable version with two 2 watt output stages.

TM340 is a deck only, without output stages.

TK320 and TS320 are two track versions of TK 341.

Grundig Model TK23, TK17L, TS19, TK23A, TK23L

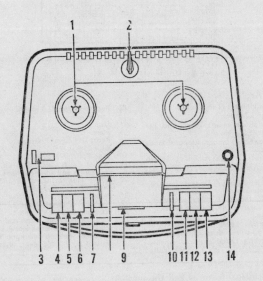

Deck Key

1 Spool Holders (5¾ in. spools).
2 Track Selector.
3 Tape Position Indicator.
4 Fast Wind Right To Left.
5 Pause Key.
6 Input Selector (depress to record from Microphone; press again to restore and record from Radio or Gram).
7 On-Off   Switch/Microphone-Diode

Recording Level (Record); Volume (Playback).
8 Sound Channel.
9 Magic Eye Signal Level Indicator.
10 Radio, P.U. Recording Level (Record) /Volume (Playback).
11 Start Key.
12 Stop Key.
13 Fast Wind Left to Right.
14 Record (press down); Superimpose (turn clockwise and press down).

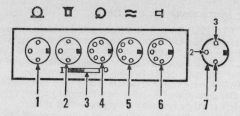

Rear Panel Key

1 Microphone Input.
2 Radio/Input (Pins 1 and 2); High Impedance Output (Pins 3 and 2).
3 Speaker On-Off Switch.
4 Gram. PU Input. (Pins 1 and 2); High Impedance Output (Pins 3 and 2).
5 Extension Speaker Output (Pins 3 and 2); MA2 Socket (Pins 3, 2, 4 and 5).
6 Output for monitor earphones SE5.
7 3-pin Socket Connexions looking on face of socket. Pin 2 must always be soldered to screen of connecting lead

Model Differences

TK17L is a De Luxe version of the TK23.

TS19, a 2-track version, has the deck mounted in a wooden case with an $8\frac{1}{2}'' \times 4\frac{5}{8}''$ speaker.

TK23A and TK23L are automatic recording versions with optional manual control. They have no mixing facilities.

Accessories

Microphone: GDM18.

Telephone Adaptor: TA3.

Earphones: STET 3 stethoscope earphones.
SE3 single earpiece with earclip.
STET stethoscope attachment for above.

Plugs: Type J3 for all input/output connexions.
Type J21 for extension speaker lead (TK18).

Connecting Leads: All Inputs and Low Impedance Outputs, SL3.
All Inputs and Outputs, SL33.
Input from Grundig Radio Diode, SL233.

Monitor Amplifier: MA2 used with STET 3 for monitoring Tracks 1 or 2 while recording Tracks 3 or 4.

Mains Voltage Adjustment

1 Take out four screws in rubber feet of case and remove cover.
2 Re-arrange links according to code below to suit your supply voltage.

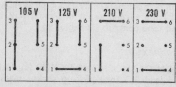

Working Instructions—in a nutshell

CONNECT TO MAINS with mains lead stowed in compartment in base.

LOAD WITH TAPE—full spool on left, empty spool on right. Wind on to right spool until metallised stop foil is through sound channel.

CONNECT INPUT from microphone, pickup, radio, etc., by inserting 3-pin plug into appropriate marked socket on rear panel.

SELECT TRACK with selector switch in centre of deck behind spools.

SWITCH ON by turning On-Off Switch/Recording Level/Volume Control on left of Magic Eye away from you until it clicks.

RECORD (a) From Microphone: depress Input Selector Key (marked MICRO). From Radio or Gram P.U.: leave key *up*. If already *down*, restore by pressing once. (b) Press and hold down Record Button on right of deck. (c) Press and hold Start Key. (d) Release Record Button and then release Start Key to start tape. (e) Press Pause Key (marked Temp. Stop). (f) Adjust Microphone Recording Level Control (left hand thumbwheel) until luminous strips just touch on strongest signals. (g) Release Pause Key to re-start tape. (h) Press Stop Key to end recording. (If you only want to pause, press Pause Key.)

WIND BACK (a) Press Fastwind Key ◀. (b) When all tape has been wound back on to left spool, press Stop Key.

PLAY BACK (a) Press Start Key. (b) Adjust volume with Volume Control (right hand thumbwheel). (c) Adjust tone with Tone Control thumbwheel on right of Magic Eye. (d) Press Stop Key to finish playback. (If you only want to pause, press Pause Key).

SWITCH OFF (a) Check that you have pressed Stop Key. (b) Turn On-Off switch (left hand thumbwheel) towards you until it clicks.

Radio Connexions

TO RECORD FROM RADIO L.S. TERMINALS (a) Push 3-pin plug on end of screened lead SL3 (supplied) into socket marked Ω on rear panel. (b) Connect red and black leads to extension speaker outlet, red to live, black to screen or chassis. Reverse black and red if hum interferes. (c) Record as above, making sure that MICRO key is *up*. (If *down*, press once to restore.)

TO RECORD FROM RADIO DIODE (a) Push 3-pin plug on end of screened lead SL33 (from your dealer) into diode socket on rear panel marked ⊔ . (b) Connect red and black leads to radio diode output, red to live and black to screen or chassis. (c) Record as above·

TO PLAY BACK THROUGH RADIO (a) Push 3-pin plug on end of screened lead SL33 into diode socket on rear panel. (b) Connect black and yellow leads to Pickup input sockets on radio receiver. (c) Switch radio to GRAM. (d) Switch off recorder speaker with switch under sockets on rear panel. (e) Play back tape as above with recorder volume control knob turned about one-third way, adjusting sound with volume and tone controls on radio.

Gramophone Connexions

TO RECORD FROM GRAMOPHONE PICKUP (a) Push 3-pin plug on end of screened lead SL3 into pick-up socket marked Ω on rear panel. (b) Connect screened lead to pick-up terminals, red to live, black to screen or chassis. (c) Start turntable and lower stylus on to disc. (d) Record as for radio L.S. above. (e) Stop and repeat from start of disc after finding correct recording level.

To Superimpose a Second Recording on the First

(a) After making first record, wind back tape and connect input for second recording. (b) Play back first record to the point where you want to start superimposing. (c) Turn Record/Superimpose button on right of deck clockwise and press down. (d) Release button at end of super-imposed section. *Note*, this method weakens the first record slightly where the second is superimposed.

To Mix Two Signals

A live recording via the microphone or a signal from a radio diode may be mixed with a recording from a radio, tape recorder or gramophone pickup this way:

(a) Plug microphone or diode leads into appropriate sockets in rear panel and radio extension speaker output or pickup into radio/pickup socket. (b) Record mixed signal using left thumbwheel to control microphone or diode input and right thumbwheel to control radio or pick-up input, so that combined signal does not close up the magic eye.

Synchronising

With the TK23 you can listen to a record previously made on one track while you make a second record to synchronise with it on the alternative track. For this you need the Monitor Amplifier MA2 connected to the Stet 3 Earphone.

(a) Make first record on Track 1 or 2 and wind back to start. (b) Connect MA2 to output socket marked ▬ on rear panel with plug on end of lead. (c) Switch track selector to alternative pair of tracks. (d) Record second item normally.

While recording you will hear the first record in the earphone, enabling you to synchronise the second recording with it. You can then hear both records together through the built in speaker by turning the track selector switch to D and playing back as usual.

Connexion to Grundig Radio

Most Grundig radios (and some other makes) have a special diode socket for connecting to a tape recorder. When you connect the diode socket of the TK40 to this socket with the screened lead SL233 (which has a 3-pin plug on each end) you can record a programme *from* and play it back *through* the radio by using the normal TK23 record/play switching arrangements.

Grundig Model TK6, TK6L

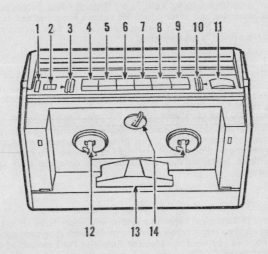

Deck Key

1 Tape Position Indicator reset.
2 Tape Position Indicator.
3 Recording Level (record)/Volume (playback).
4 Record.
5 Fastwind ◀
6 Stop.
7 Start.
8 Fastwind. ▶
9 Pause.
10 On-Off switch/Monitor speaker (record)/Tone control (playback).
11 Meter: Top scale, battery volts; Bottom scale, recording level.
12 Spool holders (4¾ in. spools).
13 Sound channel.
14 Speed Selector. Arm to left: 3¾ i.p.s. (4·75 cm/sec); arm to right: 1⅞ i.p.s (9·5 cm/sec.).

Side Panel Sockets

1 External battery socket.
2 Radio/microphone recording input, Pins 1 and 2; high impedance output Pins 2 and 3. Pickup (via socket adaptor), Pins 1 and 2.
3 Internal speaker On-Off switch.
4 Low impedance output for 5 ohm extension speaker.
5 Mains connector plug (inserted in 2-pin socket in side compartment when running off internal battery).

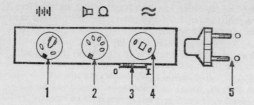

Model Differences

TK6L has a brushless motor.

Accessories

The accessories available for the TK6 include the following:

Microphone: GDM302.
Stethoscope Earphone.
Connecting Leads for Radio, Gram. P.U. and Grundig Radio.
Extension Leads for connecting to external battery.
Microphone Extension Leads.
Socket Adaptor for Gram P.U. (incorporating $\frac{1}{2}$ megohm resistor).

Working Instructions—in a nutshell

(A) Internal Battery Operation

REMOVE FRONT LID by pressing against lip on top edge. (Only necessary when preparing machine.) Stand tape recorder upright with the carrying handle at the top and the spools facing you.

LOAD WITH TAPE, full spool on left, empty spool on right. Wind on to right spool until all coloured leader tape is through sound channel. Set tape position indicator to zero.

CONNECT INPUT from microphone, pickup, radio, etc., by inserting 3-pin plug into input socket (marked ▷☐ Ω) under lid on right side of tape recorder. Replace lid, with lead running through groove provided.

SELECT SPEED with speed control on front panel.

SWITCH ON by turning right hand thumbwheel away from you until it clicks. See that needle on meter on right of press keys clears the white band on upper scale. If it does not, fit new battery (below).

RECORD. (a) Press Pause, Record and Start keys in that order. (The Pause key will stay down but you will have to hold the Record key until you press and release the Start key.) (b) Adjust recording level with left hand thumbwheel so that meter needle just touches white band (lower scale) on strongest signals. (c) Press Pause key to release and start tape. (d) To end recording press Stop key. (If you only want to pause, press Pause key once to halt tape and again to start tape.)

WIND BACK. (a) Press Fastwind key marked ◀. (b) Press Stop key as tape position indicator returns to zero to prevent tape running right off right-hand spool.

PLAY BACK. (a) Press Start key. (b) Adjust volume with left hand thumbwheel. (c) Adjust tone with right hand thumbwheel. (d) Press Stop key to finish playback. (If you only want to pause, press Pause key once to halt tape and again to start tape.)

SWITCH OFF. (a) Check that you have pressed the Stop key. (b) See that Pause key is up, if it is not, press once to release. (c) Turn right hand thumbwheel towards you until it clicks.

(B) AC Mains Operation

ADJUST MAINS VOLTAGE. The TK6 will run direct from AC mains supplies of either 220 or 110 volts. The machine as delivered is adjusted for 220v AC mains. To change to 110-volt operation do this: (a) Remove lid from back of machine. (b) Take out 2 holding screws from plastic window over mains panel. (c) Slacken screw in centre of panel and move contact arm from 220v hole to 110v hole. (d) Remove 35 mA fuse and insert 80 mA fuse. (e) Replace mains panel window and lid.

CONNECT TO MAINS. (a) Take out mains lead from right side compartment. (b) Pull 2-pin plug on end of lead out of socket in compartment and plug into AC mains socket outlet. (c) Proceed as for Internal Battery Operation (above).

Do not remove the internal batteries; removing the mains plug from the side compartment socket automatically disconnects the internal batteries.

After you switch on, the meter needle will give the same indication as for a fully charged battery.

When working from the mains the undistorted ouput steps up from 0·5 watts to 1·6 watts.

(C) External Battery Operation

The TK6 can be operated from any external DC supply voltage from 6·3v (minimum) to 11v (maximum) — e.g. a fully charged 6-volt car battery. There is no need to remove the internal batteries.

CONNECT TO BATTERY. (a) Insert plug on end of connecting lead J17 (from your Grundig dealer) into battery socket marked |||| in side compartment. (b) Connect other end of lead to external battery. (c) Proceed as for Internal Battery Operation (above).

After switching on the meter will indicate the condition of the external battery. The output power can be stepped up by removing the mains plug from its socket in the side compartment.

Internal Battery

The internal battery consists of 6, 1·5v cells. These should be Ray-o-vac Industrial Leakproof or equivalent. Fit a new set of cells this way:

(a) Remove lid from back of TK6. Insert cells, one at a time, in 2 rows of 3 in battery compartment. All cells must be fitted so that the base is to the right and the small metal top cap is to the left. (b) When fitting last pair of cells, lay plastic strap in a loop around back of recess so that it can be used to draw out the cells when they have to be changed.

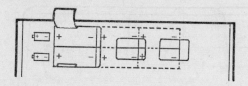

Radio Connexions

TO RECORD FROM RADIO TAPE RECORDER SOCKET (a) Push 3-pin plug on recording lead (supplied) into Radio/Micro input socket in side panel (marked ◁ ◯). (b) Connect plug at other end to radio 'Tape' socket. (c) Tune radio to required programme. (d) Record as above.

On certain Grundig radios (and a number of other makes) the record made from the tape output socket can be played back through the radio amplifier and speaker by pressing the normal playback key on the TK6. When the radio is used in this way the tape recorder speaker should be cut out with the speaker switch on the side panel.

Gramophone Connexions

TO RECORD FROM GRAMOPHONE PICKUP (a) Insert socket adaptor (from your dealer) into Radio/Micro input socket. (b) Push 3-pin plug on recording lead into socket adaptor. (c) Connect other end of lead to pickup terminals or socket. (d) Start turntable and lower pickup into groove. (e) Make test recording as above to find best recording level. (f) Wind back and make new recording from start.

Extension Speaker Connexions

(a) Connect leads from extension speaker (5 ohm) to special TK6 extension speaker plug. (b) Insert plug in extension speaker socket (marked ▭) in side panel. (c) Play back as above.

Note: If you insert the plug so that the round pin goes in to the smaller hole, you will hear the sound through the built-in speaker as well as the extension speaker. If you insert the plug the other way round, the built-in speaker will be cut out.

Grundig Model TK400

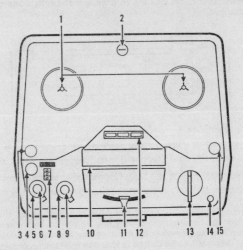

Deck Key

1 Spool Holders (5¾ in. spools).
2 Speed Control. Pressed down, 3¾; pulled up, 7½.
3 Extension speaker/Monitor amplifier socket.
4 Monitoring Earphone socket.
5 Speaker On-Off Switch.
6 Mains On-Off/Tone Control.
7 Tape position indicator.
8 Speech/Music Switch.
9 Volume.
10 Sound Channel.
11 Fast-wind: Left, Reverse; Right, Forward.
12 Track selector keys: left, Tracks 1 & 2; centre, Duoplay (both tracks at once); right, Tracks 3 & 4.
13 Record/Play back control.
14 Record safety button.
15 Microphone socket.

Socket Connexions

MICROPHONE: All inputs including microphone, radio or gramophone pick up, Pin 1; Screen, Pin 2; High impedance output, Pin 3.
MONITORING EARPHONE: Monitoring earphone, Pins 1 or 3; Screen, Pin 2.
EXTENSION SPEAKER/MONITOR AMPLIFIER: High impedance output for external amplifier, Pins 1 & 4; Screen, Pin 2, Extension Speaker, Across Pins 3 & 5.

Accessories

Microphone: GDM311
Earphones: 1. SE5 (3-pin plug to connect to monitor earphone socket).
2. SE3 (5-pin plug to connect to Extension Speaker socket or MA2).

Stethoscope Attachment: STET to fit above earphones.
Plugs: Type J3 for microphone and Monitor Earphone sockets.
Type J14 for Extension Speaker socket.
Connecting Leads: SL371R for 2-way Record/Playback connexion to radio receiver Diode socket.
SL30 for connecting to Extension Speaker socket.
Monitor Amplifier: MA2 used with STET 3 for monitoring Tracks 1 or 2 while recording Tracks 3 or 4.

Mains Voltage Adjustment

1 Unscrew rubber feet and remove base.
2 Rearrange links to code below to suit your supply voltage.

110 V	130 V	220 V	240 V
3 6	3 6	3——6	3——6
2 5	2 5	2 •5	2 •5
1 •4	1——4	1 •4	1——4

Working Instructions—in a nutshell

CONNECT TO MAINS with mains lead stowed in compartment in base.

LOAD WITH TAPE—full spool on left, empty spool on right. Wind on until metal stop foil is through sound channel.

CONNECT INPUT from microphone, pick-up, radio, etc., by inserting 3-pin plug into Microphone socket.

SELECT TRACK by pressing appropriate selector key on centre of deck: 1–2 left key; D (both tracks at once), centre key; 3–4 right key.

SWITCH ON by turning control knob on left of deck clockwise until it clicks (pilot lamp over right hand Record/Play control will light up).

SELECT SPEED with knob at back centre of deck. Pull up for $7\frac{1}{2}$; press down for $3\frac{3}{4}$. Note: Do not operate this control when recorder is switched off.

RECORD (a) Set arm of SPEECH/MUSIC control on left of deck to left (SPEECH) or right (MUSIC) according to the subject you are going to record. (b) Press red record safety button on right of deck and turn Record/Play control left to Record position. (c) To end recording restore Record/Play control to centre (Stop) position.

WIND BACK (a) Move Fastwind arm in front centre of deck to left. (b) When all tape has wound back on to left hand spool, return arm to centre position.

PLAY BACK (a) Turn Record/Play control right to Play position. (b) Adjust volume with Volume control. (c) Adjust tone with Tone control. (d) To end playback, restore Record/Play control to centre (Stop) position.

SWITCH OFF (a) Turn Mains On–Off knob on left anticlockwise until it clicks. (b) Check that Fastwind and Record/Play controls are off.

Radio Connexions

TO RECORD FROM RADIO EXT. L.S. TERMINALS (a) Use lead SL30—or SL371R with one end plug removed—and connect to an attenuating resistance network consisting of ½ megohm in series and 22 Kohm across 1 and 2. (b) Insert 3-pin plug into Microphone socket on TK400 and connect other end of lead to EXT. L.S. terminals of radio receiver with suitable end fittings. (c) Record as above; reverse connexions to radio if hum is troublesome.

TO RECORD FROM RADIO DIODE (a) If receiver is fitted with DIN-type Diode socket, connect to Microphone input of TK400 with lead SL371R. (b) Record as above. (c) If receiver has no Diode output socket one may be fitted as described on p. 54, connecting the output to a 3-pin DIN-type socket instead of to lead end 3-pin plug as shown.

TO PLAY BACK THROUGH RADIO (a) If Diode connexion has been made as above, the TK400 will automatically play back through the radio receiver power amplifier and speaker when the Record/Play control is switched to Play. (b) If there is no Diode socket on the radio receiver, the high impedence output from the Microphone socket may be connected to the Pick Up socket on the radio receiver.

Gramophone Connexions

TO RECORD FROM GRAMOPHONE PICK-UP (a) Connect pick-up to Microphone input of TK400 as for recording from Radio Ext. L.S. terminals above. (b) Start turntable and lower stylus on to disc. (c) Record as above, reversing pick-up connexions if hum is troublesome.

To Monitor While Recording

You can listen to the signal you are recording with the Grundig Earphone SE5 plugged in to the Monitor Earphone socket on the left of the deck.

Two Track Effects

You can use the two heads of the recorder to create two kinds of two track effect:

TO PLAY TWO TRACKS AT ONCE (a) Make your first recording normally. (b) Wind back to start and make your second recording on the other track by pressing down the Track key that was up while you made the first record. (c) Wind back to start, press the centre Track Selector Key, marked **D**. (d) Switch to Playback and both records will play back together.

TO SYNCHRONISE TWO TRACKS (a) Make your first recording normally with Track 1–2 key pressed. (b) Wind back to start, press Track 3–4 key. (c) Plug Monitor Amplifier MA2 into socket. (d) Plug earphone SE3 into earphone socket on MA2. (e) Switch to Record and when tape starts your first record will play back through the earphone. (f) Listen to the first record while you make the second in synchronism. (g) Play back both tracks together as above.

Grundig Model TK120

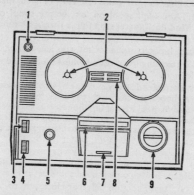

Deck Key

1 Input/output socket. Recording input (all sources), Pin 1; Screen, Pin 2 High impedance output, Pin 3.
2 Spool holders.
3 Tone control/speaker switch.
4 Recording level, Record/Volume, Playback/Mains On-Off Switch.
5 Record safety button.
6 Sound channel.
7 Magic eye recording level indicator.
8 Tape position scale.
9 Function selector—(Functions, reading clockwise: Fastwind forward, Start, Pause, Stop, Fastwind reverse).

Accessories

Microphone: GDM311
Plugs: Type J3 for all inputs and high impedance output.
Connecting Leads: SL32R supplied.
SL371R.
SL30.

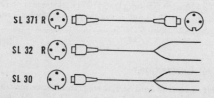

Mains Voltage Adjustment

1 Unscrew retaining screw in cover plate on back of case and remove plate.
2 Pull out voltage selector knob until you can turn it and set it to your supply voltage.
3 Push in selector to lock and replace cover plate.

Q   xlix

Working Instructions—in a nutshell

CONNECT TO MAINS with mains lead stowed in compartment in base.

LOAD WITH TAPE—full spool on left, empty spool on right. Wind forward by hand until metal stop foil is through sound channel.

CONNECT INPUT from microphone, radio, pick-up, etc., by inserting 3-pin plug into Input/Output socket, top left corner of deck.

SWITCH ON by turning front thumbwheel on left of deck towards you until it clicks.

RECORD (a) Press Record Safety button on left of deck and turn Function Selector on right of deck clockwise to Pause. (b) Adjust recording level with front thumbwheel so that luminous strips on recording level indicator just touch on the strongest signals. (c) Turn Selector on to Start. (d) To end recording restore Selector to Stop position. (If you only want to pause for a moment or two, just turn Selector as far as Pause position.)

REWIND (a) Turn Selector anticlockwise to Fast Rewind position. (b) When all tape is back on left hand spool return Selector to Stop.

PLAY BACK (a) Turn Function Selector clockwise to Start. (b) Adjust volume with front thumbwheel and tone with rear thumbwheel. (c) To end playback, turn Selector back to Stop position. (If you only want to pause, just turn Selector as far as Pause position.)

SWITCH OFF (a) Return Function Selector to Stop position. (b) Switch off mains by turning front thumbwheel away from you till it clicks.

Radio Connexions

TO RECORD FROM RADIO EXT. L.S. TERMINALS. (a) Push 3-pin plug on end of lead SL32R (supplied) into Input/Output socket and connect free ends to extension speaker terminals on radio. (b) Record normally, reversing speaker connexions if hum is troublesome.

TO RECORD AND PLAY BACK THROUGH DIODE SOCKET (a) If your radio has a special Diode socket, connect this to the Input/Output socket on TK120 with Grundig lead SL371R. If there is no Diode socket, you can adapt lead SL30 and a resistance network to make a 2-way connexion to your radio circuit.

Gramophone Connexions

(a) Connect free ends of lead SL32R to gramophone pick-up and insert 3-pin plug into Input/Output socket on TK120. (b) Start turntable, lower stylus on to disc, and record normally.

1

Grundig Model C100

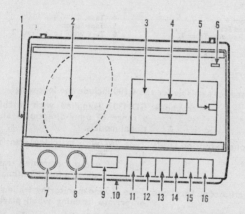

Deck Key

1 (On side) Input/Output Sockets (see separate diagram).
2 Speaker Grille.
3 Casette cover.
4 Tape ObservationWindow.
5 Casette Cover Catch.
6 Mains Unit Pilot Light.
7 Recording Level (record)/Volume (play back).
8 Speaker Switch/Tone Control.

9 Meter: Top Scale, Recording Level Bottom Scale, Battery Volts.
10 Pull-out carrying handle.
11 Record.
12 Fastwind.
13 Stop.
14 Start.
15 Fastwind.
16 Pause.

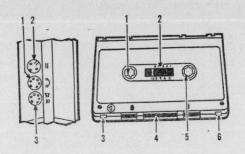

Input/Output Socket

1 Monitor Earphone.
2 External Battery (6.3v to 9v).

3 Microphone/Radio/Pickup Recording Input.

Cassette

1 Tape Spool.
2 Tape Used Scale.
3 Tape Erase Lock, Track 1.

4 Tape.
5 Tape Spool.
6 Tape Erase Lock, Track 2.

Accessories

The accessories available for the C100 include the following:

Microphone: GDM302 (supplied with machine) and a range of omni-directional and directional models.

Microphone Extension Leads: Type 267 with matching transformer 5, 10 and 15m long.
Type 268 without matching transformer to extend above a further 10m.

Stethoscope Earphone: Type 210 for monitoring while recording or private listening when playing back.

Connecting Leads: Type 237, 2.5m long, with 3-pole plug for connecting to radio or second recorder.

Socket Adaptor: Type 293, for connection to record player.

Car Connexion: Type 385, for connecting to 6v car battery.
Type 386, as above for 12v car battery.

Mains Unit to replace batteries and run tape recorder off AC mains 50/60 cycles, 110/220v.

lii

Working Instructions—in a nutshell

(A) Internal Battery Operations

LOAD WITH TAPE CASSETTE (a) Press catch on tape cassette compartment cover to release cover. (b) Insert casette with tape aperture towards handle of recorder. (If necessary press down either fast wind key for a second to turn spindles to engage in spool centres.) (c) Replace cassette cover.

CONNECT INPUT from microphone, radio, pickup etc. by inserting 3-pole plug in appropriate socket on side panel.

RECORD (a) Press Pause key, hold down Record key and press Start key. Record and Start keys will now stay down. (b) Adjust recording level with left hand control so that meter needle just touches red band on upper scale on strongest signals (c) Press Pause key to start recording (d) To end recording press Stop key. (If you only want to pause, press Pause key once to halt tape and again to start tape.)

WIND BACK (a) Press Fastwind key marked ◀◀. (b) When all the tape has wound back to the start, press Stop key.

PLAY BACK (a) Turn right hand knob clockwise to switch on internal speaker. (b) Press Start key (c) Adjust volume with left hand knob (d) Adjust tone with right hand knob. (e) Press Stop key to finish play back. (If you only want to pause, press Pause key once to stop tape and again to start.)

(B) AC Mains Operation

The C100 can be operated off normal 110 or 220 volt AC mains by using the special Power Pack. First you remove the battery container which is housed in the back of the case. To do this you slide back the catch in the centre and ease the container out by the lip on the bottom edge. The Power Pack fits into the same space and is connected to the mains by its lead and plug. The voltage must be adjusted to the local supply. The pack is then switched on with its independent switch and operation from that point is exactly as for battery operation.

External Battery Operation

The C100 can be operated from an external battery—e.g. a car battery—by simply connecting it to the battery by a suitable connecting lead plugged in to the external battery socket on the side panel. There are two leads available: Type 385, for use with a 6-volt battery and Type 386 for use with a 12-volt battery. There is no need to remove the internal battery; it is automatically switched out of circuit when the plug on the connecting lead is inserted in the socket.

Internal Battery

The internal battery consists of 6, 1.5v cells. These should be of a heavy duty, leakproof type. To fit a new set of cells (when the indicator needle fails to move out of the red band when recording or playing back):

liii

(a) Remove battery container as described above. (b) Remove old cells. A hole in the back of the container lets you push out the first cell and the rest will follow easily. (c) Insert new cells making sure they lie as shown on the diagrams moulded on the case. (d) Fit container into recorder and press the Start key. If the meter needle clears the lower red band the battery is satisfactory.

Radio Connexions

TO RECORD FROM RADIO TAPE RECORDER SOCKET (a) Connect recorder to radio by connecting lead Type 237 plugged in to Tape socket on radio set and Radio/Microphone socket on recorder. (b) Tune radio to required programme. (c) Record as above.

On certain Grundig radios (and a number of other makes) you can play back the record you have just made through the radio amplifier and speaker by simply switching the recorder to play back. When you do this, turn the tape recorder volume control to maximum, switch off the internal speaker, and regulate the volume from the speaker with the radio volume control.

Other Equipment

You can also record from a second tape recorder or a record player by connecting it to the C100 with connecting lead Type 237. The plugs on either end of this lead can be plugged in to the tape recorder but it may be necessary to change over the second plug from pins to sockets to suit the other equipment. This is done with a socket adaptor Type 239.

Cassette

The tape cassette conforms to the international standard DC design used for pre-recorded tapes. When loaded with pre-recorded tape the cassette is specially modified to prevent the record key from being accidentally pressed and erasing the record. A supply of moulded stops is sent out with each C100 recorder. The erase action can be blocked by pressing one of these stops into the hole in the cassette below the moulded track number. It will then prevent accidental erasure of that track as long as it is left in position. It can be removed by prizing it out with a sharp point.

Grundig Models TK125, TK140 & TK145

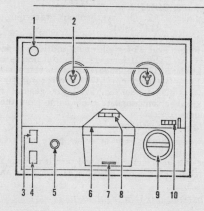

Deck Key

1 All Inputs (Pins 1 and 2) and High Impedance Output (Pins 3 and 2).
2 Spool Holders (5¾ in. spools).
3 Tone Control/Speaker Switch.
4 Recording Level, Record/Volume, Playback/Mains On-Off Switch.
5 TK140: Record Safety Button only. TK125: Record Safety Button incorporating Automatic/Manual Recording Switch with Speech-Music position on Automatic and extra Trick (Superimpose) position on Manual.
6 Sound Channel.
7 Magic Eye Recording Level Indicator.
8 Track Selector (TK140 only).
9 Function Selector. Functions, Reading Clockwise: Fastwind Forward, Start, Pause, Stop, Fastwind Back.)
10 Tape Position Indicator.

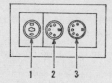

Rear Panel Key

1 Extension Speaker: With round pin of plug J21 in top socket, internal speaker is cut out; with round pin in bottom socket, both speakers are connected.

2 Monitor Headphone, Pins 2 and 3.

3 Monitor Amplifier. (TK140 only.)

Model Differences

TK125 is a 2-track model with both Automatic and Manual recording level control and Trick (superimpose) switch.

TK140 is a 4-track model with facility for synchronous recording and tape cleaner.

TK145 is a 4-track Auto/Manual model with automatic end-of-tape stop.

Accessories

Microphone: GDM312.

Monitor Amplifier: MA2 for synchronous recording (TK140 and TK145 only).

Earphone: Type SE3.

Plugs: Type J3 for all inputs and high impedance output. Type J21 for switched extension speaker.

Connecting Leads: SL32R (supplied) for general recording (incorporates attenuator), SL371R for record/play connexion to Diode socket, SL30 for general recording (no attenuator).

Mains Voltage Adjustment

1 Unscrew retaining screw in cover plate on back of case and remove plate.

2 Pull out voltage selector knob and turn it to your supply voltage.

3 Push back selector knob and replace cover plate and screw.

Working Instructions—in a nutshell

CONNECT TO MAINS with mains lead stowed in compartment in base.

LOAD WITH TAPE, full spool on left, empty spool on right. Wind forward by hand until metal stop foil is through sound channel.

CONNECT INPUT from microphone, radio, pickup, etc. by inserting 3-pin DIN plug into Input/Output socket, top left corner of deck.

SELECT TRACK (TK140 and TK145 only) by pressing down appropriate track key in centre of deck.

SWITCH ON by turning front thumbwheel on left of deck towards you until it clicks.

RECORD (a) TK140: Press record safety button and turn function selector to Pause. TK125 and TK145: for automatic recording, turn record safety button to appropriate Speech or Music position; for manual recording, turn to Level position. Then turn function selector to Pause. (b) Adjust recording level with front thumbwheel so that luminous strips on recording level indicator just touch on strongest signals. Ignore this step with TK125 and TK145 set for automatic recording. (c) Turn function selector knob to Start. (d) To end recording, restore function selector to Stop. (If you only want to pause for a moment, turn only as far as Pause.)

REWIND (a) Turn function selector anticlockwise to Fast Rewind position. (b) When all tape is back on lefthand spool, return to Stop.

lvi

PLAY BACK (a) Turn function selector to Start. (b) Adjust volume with front thumbwheel, tone with rear thumbwheel. (c) To end playback, turn selector to Stop. (If you only want to pause, turn only as far as Pause.)

SWITCH OFF (a) Return function selector to Stop. (b) Switch off mains by turning front thumbwheel away from you till it clicks.

Superimposing (TK125 only)

TO SUPERIMPOSE A SECOND RECORDING OVER THE FIRST (a) Adjust level for second recording in the normal way. (b) Switch to Playback and play tape to point where you want to superimpose. (c) Turn selector to Pause. (d) Turn record safety button to Trick and hold down. (e) Turn selector to Start. (f) Record normally and release record safety button at end of superimposed item.

Two Track Effects (TK140 and TK145 only)

TO PLAY BACK TWO TRACKS AT ONCE (a) Make two separate recordings on the same section of tape, on Tracks 1 and 3, or 2 and 4. (b) Wind tape back to start. (c) Press track selector key D and play back normally. Both recordings will play together.

TO SYNCHRONISE TWO RECORDINGS (a) Make first recording with Track 1–2 key pressed and wind back to start. (b) Connect earphone SE3 to monitor amplifier socket on rear panel through monitor amplifier MA2. (c) Press Track 3–4 key and prepare to make second recording. (d) Start tape and make second recording as you listen to first in earphone, synchronising the second with the first.

Note: It does not matter whether the two tracks are recorded in the above order or the other way round.

Radio Connexions

TO RECORD FROM RADIO EXT L.S. TERMINALS (a) Push 3-pin plug on end of lead SL32R (supplied) into Input/Output socket and connect free ends to extension speaker terminals on radio. (b) Record normally, reversing speaker connexions if hum is troublesome.

TO RECORD AND PLAY BACK THROUGH DIODE SOCKET (a) If your radio has a special Diode socket, connect this to the Input/Output socket with Grundig lead SL371R. If there is no Diode socket, the diagram on p. 54 will show you how to make the necessary modifications to your radio so that you can record and play back using lead SL30.

Gramophone Connexions

TO RECORD FROM PICKUP (a) Connect free ends of lead SL32R to pickup and insert 3-pin plug into Input/Output socket on tape recorder. (b) Start turntable, lower stylus onto disc and record normally.

Grundig Model TK2200

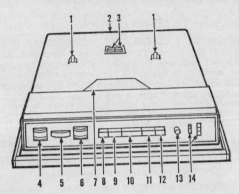

Deck Key

1 Spool Holders (5″ spools).
2 Battery and Power Pack compartment (in base).
3 Speed Selector (3¾ and 1⅞).
4 Volume.
5 Recording level/Battery condition indicator.
6 Recording level (bottom ring) Tone (top ring).
7 Sound Channel.
8 Fast rewind.
9 Pause.
10 Stop.
11 Start.
12 Fast forward wind.
13 Record interlock.
14 Tape position indicator.

Socket Panel Key

(Note: With the recorder standing upright and spools facing you the socket panel is on the right hand side.)

1 Universal input/output: recording from microphone or external equipment, Pins 1 & 2; playing back through external equipment, Pins 3 & 2. Screen, Pin 2. (If the signal you want to record is too strong you will have to use a lead incorporating an attenuator—e.g. Grundig lead 293.)

2 Connexion to external D.C. power supply of 6 to 10 volts. Inserting plug automatically disconnects internal battery.

3 External speaker or monitor earphone type 340. Inserting plug automatically disconnects built in speaker.

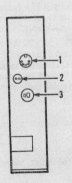

Accessories

Dynamic Microphones: GDM301S Cardiod (directional) microphone with remote control switch.
GDM305 Omnidirectional microphone with remote control switch.

Stethoscope Earphone: Type 340 for monitoring and personal listening.

Connecting Leads: Type 237 for recording and playing back via the tape socket on a radio.
Type 293 as above but fitted with an attenuator to reduce the strength of the signal when connected to certain gramophones or a second tape recorder.

Car Connexions: For connecting to car battery. Type 381 (6.3V) and Type 386 (12V).

The TK2200 may also be used with many other accessories listed in the Accessories section.

Working Instructions—in a nutshell

REMOVE FRONT LID by pressing buttons at each side and pushing cover towards hinges.

LOAD WITH TAPE with base of recorder away from you, fit full spool on left holder and empty spool on right.

CONNECT INPUT with 3-pin DIN type plug to universal input/output socket on side of recorder.

SELECT SPEED by pressing $1\frac{7}{8}$ or $3\frac{3}{4}$ i.p.s. speed selector key on deck between spools as required.

SWITCH ON You switch on automatically when you press the start key or either fastwind key.

RECORD (a) Press Pause, Record and Start Keys in that order. (b) Release pressure on keys in same order. (c) Adjust recording level with bottom section of right hand thumbwheel so that indicator needle just reaches white/red dividing line on loudest signals. (d) Press Pause Key to release and start tape. (e) Press Stop Key to end recording. If you only want to pause, use Pause Key.

WIND BACK (a) Press fast rewind key on left of control panel. (b) Press Stop Key when all tape is back on left hand spool. (Use the tape position indicator to avoid running the end of the tape right off the spool.)

Note: The TK2200 has an automatic stop which switches off the recorder if the end of the tape runs off the spool or if the tape breaks.

PLAY BACK (a) Press Start Key. (b) Adjust volume with left hand thumbwheel and tone with top section of right hand thumbwheel. (c) Press Stop Key to finish play back. If you only want to pause, use Pause Key.

SWITCH OFF (a) Press Stop Key. (b) If Pause Key is down, press once to restore.

(A) Internal Battery Operation

(a) Release lid of battery compartment in base by pressing spring catch. (b) Fit 6, 1.5v leakproof drycells in compartment arranged in accordance with diagram printed inside. (c) Slide Mains/Battery switch in compartment to BATT. (d) Replace lid. (e) Check battery condition by pressing Pause and Start Keys. If pointer on indicator fails to move into white section of scale the battery should be replaced with a new one.

(B) External Battery Operation

Any external direct current supply between 6 and 10 volts may be used to replace the internal battery. If the external supply is connected to the power input socket via a twin lead and 2-pin plug the internal battery is automatically disconnected and the recorder can be operated normally using the external power supply.

(C) Mains Operation

The internal battery may be replaced by the Grundig TN12 mains unit. This can be done two ways:

(1) The mains unit may be plugged in to the power input socket like the external battery above).

(2) The mains unit may be fitted into the recess provided at the end of the battery compartment. It is then connected to the recorder by the press stud connectors on the wander lead inside the recess. In this case the slide switch must be set to MAINS.

Note: If you intend to run the recorder off an alternative power supply for any length of time you should remove the internal battery so that no harm will be done if it should develop a leak. (This is rare but not impossible with modern drycells).

Radio/Gram Connexions

The TK2200 can be connected by lead Type 237 to the tape socket of a radio or radiogram. The recorder will then record whatever is being played on the equipment—even when the volume control of the radio or gram is turned down. It will also play back through the amplifier and speaker of the external equipment, with the same connecting lead. The general procedure for recording and playing back through external equipment is given in Chapter 3.

Extension Speaker

The TK2200 will play back through an extension speaker (approx. 4 ohms) or, for private listening, through a headphone (1.8 Kohm)—e.g. Grundig Type 340. These should be connected to the extension speaker socket. Plugging in the external unit will automatically disconnect the internal speaker.

Monitoring

You can listen on the internal speaker or on an extension speaker or headphone to the signal you are recording by turning up the normal volume control. If you want to monitor a recording from the microphone, however, you should always use headphones connected to the extension speaker socket to prevent the signal from feeding back into the microphone and starting a howl.

Remote Control

Two of the Grundig microphones for use with the TK2200—GDM 301S and GDM305—incorporate a switch for starting and stopping the recorder without affecting the normal controls.

Grundig Model C200

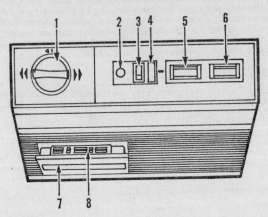

Deck Key

1 Function Selector. Functions: reading clockwise from Stop position (bar parallel to front of cabinet): Pause; Start tape.
To release cassette, turn anticlockwise against spring.
To fastwind forward, press to right.
To fastwind back, press to left.

2 Record interlock button.
3 Recording level meter, record/battery condition indicator, playback.
4 Recording level control.
5 Volume control, playback.
6 Tone control, playback.
7 Cassette holder cover.
8 Tape cassette.

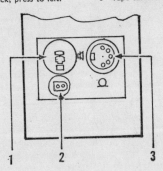

Socket Key

1 Output for monitor earphone, type 340, 1.8 K ohm or external speaker, 5 ohm.

2 Input for external d.c. power supply, 5–7.5 volts.

3 Universal input/output: Input for microphone/radio/pick up, 0.2–15 mV at approx. 7K ohms or 65 mV–4.8V at approx. 2.2 M ohms. Output for external amplifier, 600 mV at 18 K ohms.

Accessories

The accessories available for the C200 include:

> Tape Cassettes, C60 (playing time 2 x 30 min.)
> C90 (playing time 2 x 45 min.)
> C120 (playing time 2 x 60 min.)

Dynamic Microphones: GDM303 with remote control.
GDM301 and GDM304.

Earphone Type 340 for monitoring and private listening.

Car Battery Connexion Cable, Type 381, with noise suppressor (6-volt battery). Adaptor Type 386 for 12 volt battery.

Mains Unit TN12 Universal for running C200 off mains.

Socket Adaptor Type 293 for connecting C200 to a record player.

Note: A support, Type 473 will be available for fitting the C200 into a car and automatically connecting it to the car radio and its power supply. Two connecting Kits: R, Type 474 and L, Type 475 are also in preparation for recording and playing back via the car radio.

Working Instructions—in a nutshell

FIT BATTERY (a) Release cover by pressing two buttons in base. (b) Assemble 5, U11 batteries in plastic tube, following polarity diagram moulded in to the underside of the lid. (c) Replace cover.

LOAD WITH TAPE CASSETTE (a) Turn function selector anticlockwise to open lid of cassette compartment. (b) Insert cassette with exposed tape at top and the side to be played facing the lid of the compartment. (c) Press cassette compartment lid in to close.

CONNECT INPUT from microphone, pickup, radio etc. with 3-pin DIN plug to input/output socket on right side of case.

RECORD (a) Press record button and turn function selector to Pause. (b) Adjust recording level with recording level thumbwheel, so that needle just reaches red line on loudest signals. (c) To start recording turn function selector to Start. (d) To end recording turn function selector to Stop. (If you only want to pause, turn to Pause.)

REWIND (a) Press function selector to left and hold. (b) When all tape has run back, release function selector.

PLAY (a) Turn function selector to Start. (b) Adjust volume with Volume thumbwheel. (c) Adjust tone with Tone thumbwheel. (d) To end playback turn function selector to Stop. (If you only want to pause, turn to Pause.)

SWITCH OFF. See that function selector is at Stop.

REMOVE CASSETTE (a) Turn function selector anticlockwise until lid of cassette compartment springs open. (b) Remove cassette.

Alternative Power Sources

The C200 may be operated from any external d.c. source of 5 to 7.5 volts, preferably from the Grundig transistorised power pack, Type TN12, plugged into the external power supply socket.

State of Battery

The meter on top of the recorder indicates the state of the internal battery or of any external power supply to which the recorder is connected. The needle should lie in the green section when the function selector is at Pause or Start. If the needle lies below the red line fit a new battery.

Record/Play Connexions

The C200 will record from and play back through a radio when connected to it with lead Type 237. The volume control of the C200 may be turned down to mute the built in speaker without affecting the sound from the radio speaker. It will also play back through an extension speaker plugged in to the extension speaker output socket. In this case the volume and tone controls of the C200 remain operative.

Monitoring

When recording either through the microphone or universal input/output socket, the signal may be monitored with earphone Type 340 plugged in to the extension speaker socket.

Grundig Model TK245

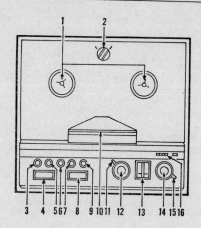

Deck Key

1 Spool Holders.
2 Speed selector/On-off and pilot lamp switch.
3 Multiplay.
4 Fast rewind.
5 Recording level (manual control).
6 Record interlock button.
7 Pause.
8 Stop.
9 Start.
10 Sound Channel.

11 Track & function selector: Tracks 1–2, 3–4; stereo. Record and Playback; Parallel Play 1 & 3 or 2 & 4.
12 Cross recording for Multiplay and tone control, for playback.
13 Recording level meter.
14 Recording level control, record/Volume control, playback.
15 Input selector: Microphone; radio or record player.
16 Tape position indicator.

Socket Panel

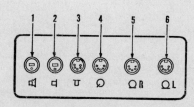

Socket Key

1 Output: External speaker (approx. 5 ohm). Insert plug with round pin at top to switch off internal speaker. With record pin below, both speakers will operate.

2 Output: For mono monitoring on single earphone, Type 210 or monitoring right and left hand stereo channels or two independently recorded tracks on two 210 earphones via distributor lead type 276.

3 Input/output: For recording from and playing back through a radio receiver,

stereo or mono and other recording and playback operations with external equipment, using lead Type 24 .

4 Input: For recording from record player, or second tape recorder, mono or stereo, using lead Type 242.

5 Input: Right hand channel: For recording from mono mcrophone.

6 Input; mono and stereo: For recording from mono microphone (left hand channel only) or stereo microphone (both channels.)

Model Differences

Accessories

Microphones: Dynamic. Types GDM317, GDM321, GDM322 and GDM330.

Tripod support: S15 with Flexible pillar MSH 20.

Extension leads: Type 267, 5, 10 or 15 metres long, with matching transformer.
Type 268, 10 metres long, without matching transformer.

Connexion Leads: Type 237, 2½ metres long, with 2, 3-pin plugs for connecting to external mono equipment.
Type 242, 2 metres long, for connecting to external stereo equipment (including stereo mixer 422).
Monitoring Earphone: Type 210.

Stereo Mixer: Type 422 with two inputs into independent transistor amplifiers for recording two channels via lead 242 and radio socket.

Telephone Adaptor: Type 244 U for recording both sides of a telephone conversation.

Mains Voltage Adjustment

Check the figure indicated by the arrow on the mains voltage adjuster on rear panel of recorder If different from your local electricity supply voltage, use coin to turn adjuster to correct figure

Working Instructions—in a nutshell

CONNECT TO MAINS with 3-core mains lead stowed in compartment in base.

LOAD WITH TAPE, full spool on left, empty spool on right. Wind on until metal stop foil is through the sound channel.

CONNECT INPUT from mono and stereo sources with lead and plug via relevant socket From 2 independent sources via lead 242 and radio socket.

SELECT INPUT by moving input selector arm to ⏚ 𝒪 if recording through radio socket or 𝒪 if recording from microphone.

SELECT TRACK by setting arm of track switch to 1–2 or 3–4 (for mono recording and playback) or S (for stereo recording and playback).

SELECT SPEED (AND SWITCH ON) by moving speed switch from 0 to 7½ or 3¾. The meter will light up to show that the circuit is ready to operate.

RECORD (Manual) (a) Press Manual and Recording buttons. (b) Adjust level control until the operative indicator (mono) or both indicators (stereo) just rises to 7 on strongest signals. (c) Press start key to start recording (d) Press stop key to end recording. If you only want to pause, use pause buttons. (Automatic) For automatic recording, omit steps (a) and (b) above.

WIND BACK (a) Move fastwind control to left. (b) When all tape has wound on to left spool the automatic stop operates arresting the tape and restoring fastwind control.

PLAYBACK (a) Press start button. (b) Adjust volume and tone with appropriate controls. (c) To end playback press stop key. If you only want to pause, use pause button.

SWITCH OFF (a) Press stop button. (b) Turn speed switch to centre zero.

Radio/Gram Connexions

This recorder will make records from a radio receiver or radiogram and play back the recorded tape through the amplifier(s) and speaker(s) of the same equipment using a single connecting lead plugged in to the tape socket of the external equipment and the diode socket of the recorder (Use lead Type 237 for mono and Type 242 for stereo.)

Stereo Playback

The recorder is designed to record both stereo channels but as there is only a single internal power amplifier and speaker assembly it just plays back the right hand channel. However, there are two ways of playing back both channels for true stereo reproduction:

1. The left hand channel can be reproduced through the amplifier and speaker of a radio by connecting the diode sockets of the recorder to the tape socket of the radio with Grundig connecting cable Type 242. As the right hand channel will then play through the built in speaker, the tape recorder should stand on the right of the radio.

2. Both channels may be played through stereo reproduction equipment by connecting the diode socket of the recorder to the tape socket of the stereo equipment. The recorder volume control should be turned down to mute the built in speakers.

Two Track Effects

The two heads of the TK245 de luxe make it possible to produce a number of special effects by combining separately recorded mono recordings:

Combining Two Tracks

To play two tracks, separately recorded on 1 or 2 and 3 or 4 of the same tape—e.g. music and the spoken word—turn track selector to D and proceed as for normal playback.

Synchronising Two Tracks

(a) Make your first mono recording in the normal way. (b) Wind back to start. (c) Change track selector switch to alternative track. (d) Connect earphone to earphone output socket with lead Type 276, through the yellow adaptor plug. (e) Switch to manual recording and make a preliminary check to get the level of the second recording right. (f) Start the tape and listen to the first recording in the earphone while you make the second in synchronism. (g) Wind back to start and play the combined recording with the track selector switched to D.

Adding Further Recordings

When you make a synchronised recording as above, if you press the Multiplay Key, your first recording will record on the new track together with the second recording. You can then hear the combined recording by leaving the selector at the track you have just recorded and playing back in the normal way. To add a third recording, you simply repeat the procedure but change over the track selector. You will then combine your third recording with the combined first and second recordings on to the new track. The process can be continued as often as you please.

Note: The normal recording level control sets the level of the recording being added; the Multiplay control sets the level at which the previous recordings are transferred to the new track. It is best to make a preliminary run to set the Multiplay level before adding the next recording.

Monitoring

You can listen to the signal you are recording by connecting earphone Type 210 to the earphone socket via adaptor lead 276. The yellow socket connects to the right hand stereo channel and the red socket to the left hand channel.

Grundig Model 247 de luxe

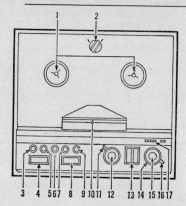

Deck Key

1 Spool Holders.
2 Speed selector/On-off and pilot lamp switch.
3 Multiplay.
4 Fast rewind.
5 Recording level (manual control).
6 Record interlock button.
7 Pause.
8 Stop.
9 Start.
10 Sound Channel.
11 Track & function selector: Tracks 1–2, 3–4; stereo. Record and Playback; Parallel Pay 1 & 3 and 2 & 4.

12 Monitor control (record)/Volume & balance (playback). Pull up top button to use as balance control.
13 Recording level meter.
14 Cross recording for Multiplay and tone control, playback.
15 Recording level control, record/ Volume control for playback.
16 Input selector: Microphone; radio or record player.
17 Tape position indicator.

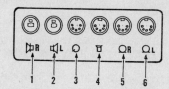

Socket Key

1 Output: External speaker right hand channel (approx. 5 ohm). When plug is inserted, internal speaker is switched off.
2 Output: External speaker, left hand channel (approx. 5 ohm).
3 Input: For recording from record player, or second tape recorder, mono or stereo, using lead Type 242.
4 Input/output: For recording from and playing back through a radio receiver, stereo or mono and other recording and playback operations with external equipment, using lead Type 242.
5 Input: Right hand channel: For recording from mono microphone.
6 Input; mono and stereo: For recording from mono microphone (left hand channel only) or stereo microphone (both channels.)

Accessories

Among the accessories available for the TK247 de luxe are:

Dynamic Microphones: GDM312U Omnidirectional stick type.
GDM317U Cardioid stick type, with tripod stand.
GDSM330U Twin cardioid with tripod stand, for stereo recording.

Microphone Extension Leads: Type 267 with matching transformer.
Type 268 without matching transformer.

Earphone: Type 340, monitoring, single channel.
Type 211 high quality stereo for personal listening.

Stereo Mixer: Type 422, transistorised, 4-channel.

Connecting Leads: Type 237 for mono recording and playback with external equipment.
Type 242 for stereo recording and playback with external equipment.

The TK247 de luxe may also be used with many other accessories listed in the Accessories section.

Mains Voltage Adjustment

The mains voltage is adjusted by turning the red disc in the back of the recorder. See that the arrow is pointing to your local supply voltage. If necessary, insert a coin in the slot and turn to indicate the correct figure. Your dealer can also adjust the recorder to operate off a 50 or 60 Hz supply frequency.

Working Instructions—in a nutshell

CONNECT TO MAINS with 3 core lead in base compartment fitted with suitable plug.

LOAD WITH TAPE full spool on left, empty spool on right.

CONNECT INPUT for microphone, radio or pickup by inserting suitable plug into appropriate socket on rear socket panel. (For mono) recording with a single microphone, insert plug in Left channel microphone socket.

SELECT INPUT by turning selector arm on right of deck to microphone or radio.

SELECT SPEED by turning speed selector to $7\frac{1}{2}$ or $3\frac{3}{4}$ as required.

SELECT TRACK by turning arm of track selector to 1–2 or 3–4 for mono or S for stereo.

SWITCH ON You switch on automatically when you select the speed.

RECORD (a) Turn volume control (on left) anticlockwise unless you want to monitor the signal you are recording through the internal speakers (b) Press recording button (in middle of row, marked ▲).

(c) Adjust recording level with lower section of right hand control knob. (d) To start recording press start button. (e) To end recording press stop key. If you only want to pause, use Pause Button.

WIND BACK (a) Move fastwind slider to left. (b) When all tape has wound back, press stop key.

PLAY BACK (a) For mono recordings turn track selector to number of recorded track, for a stereo recording, turn selector to S. (b) To start playback, press start button. (c) Adjust balance between right and left channel built in speakers by pulling up and turning upper section of left hand control knob. Turning to left increases sound from left speaker and vice versa. When correct balance results, push down and operate the whole knob to control the overall volume. (d) Adjust tone with upper section of right hand control knob. (e) To end playback, press Stop Key.

SWITCH OFF (a) Press Stop Key to return all controls to off. (b) Turn speed selector to centre position.

Radio/Gram Connexions

The TK247 de luxe will record from a mono or stereo radio receiver and play back the recorded tape through the amplifier(s) and speaker(s) of the same equipment, using a single lead (Type 237 for mono and Type 242 for stereo) These leads carry suitable multipin plugs at each end for connecting to the tape recorder diode (radio) or pickup sockets and the tape socket of the external equipment.

The recorder will also play back stereo tapes through a stereo hi-fi reproducer (using lead 242) or through right and left channel extension speakers plugged in to the extension speaker sockets. When playing back through external equipment plugged in to the diode or pickup sockets, the internal speakers can be turned right down. When playing through extension speakers inserting the plugs automatically switches off the internal speakers.

Two Track Effects

The two recording and playback sections of this stereo recorder can be operated independently to create a number of special effects.

Combining Two Tracks

(a) Make a mono record in the normal way (b) Rewind tape and change track selector to alternative tracks (c) Make a second mono record on the new track. (d) Rewind tape. (e) Turn track selector to D. (f) Press start key and both tracks will play back together.

Synchronising Two Tracks

(a) Make a mono record in the normal way. (b) Rewind tape and change track selector to alternative tracks. (c) Connect earphone Type 340 to Left channel extension speaker socket. (d) Make your second record and as you do so, listen to your first record in the

lxxi

headphone and synchronise your new recording with it. (If you wish you can connect a second monitor headphone to the other extension speaker socket for a separate recording engineer). (e) Rewind tape and play back synchronised record with track selector switched to D.

Adding Further Recordings

The multiplay controls let you build up your recordings as you go along—e.g. to turn a duet into a trio and then into a quartet. This is how you do it: (a) Make your first record normally and wind back to start. (b) Turn track selector to alternative tracks. (c) Press Multiplay (the Play Key will also depress) and Record buttons. (d) Turn normal Recording Level control back to zero. (e) Press Start button and adjust level of cross recording with multiplay (upper) section of control as for normal recording. (f) Rewind tape. (g) Adjust recording level for second recording using normal Level control. (h) Start tape and make second recording. Do not move Multiplay control.

The first and second recordings are now combined on the second track. You can now switch to the other track and add them both to a third recording and so on by repeating the above procedure.

Monitoring

You can hear the signal you are recording on the internal speakers by turning up the volume control. If you are recording from a microphone in the same room, this may set up a howl. In this case it is best to monitor the signal with a headphone Type 340 plugged in to an extension speaker socket. For stereo records you can use either two of these headphones or one of the special Grundig stereo headphones. When you plug in to the extension speaker socket in this way you automatically cut out the internal speakers.

Grundig Accessories

MICROPHONES   *Early Models no longer in Production*

	Type	Directional Characteristics	Impedance	Sensitivity mV/microbar
GCM 1 and 3	Con.	Omni	High	2·5
GDM 111	D	Cardioid	High	1·75
GDM 32/52	D	Omni	High	0·9
GDM 5	D	Omni	Low	0·03
GRM 1/2	R	Fig. 8	Low	0·025
GRM 12/32	R	Fig. 8	High	0·7
GXM 1	Cry	Omni	High	1·5
GM1 /1L	D	Omni	Low	—

Current Models

	Type	Directional Characteristics	Impedance	Sensitivity mV/microbar
GDM 12/18 */311/19	D	Omni	200	0·15
			55 k	2·2
GDSM 202	D	2 x Fig. 8	200	0·1
			55 k	1·5
GDM 121	D	Omni	200	0·3
			40 k	3·0
GBM 125	R	Cardioid	200	0·1
			200 k	3·0
GDM 300	D	Omni	4,000	0·35
GDM 302	D	Omni	4,000	0·35
GDM 304	D (stick)	Omni	4,000	0·35
GDM 311	D	Omni	200	0·35
			55 k	
GDM 312	D (stick)	Omni	200	1·5
			55 k	

Microphone types are abbreviated as follows:—

Con. = Condenser; D = Dynamic; R = Ribbon; Cry. = Crystal.
* If the plastic microphone container for this model is marked GDM 18—Hi, the microphone has a high impedance output only.

STETHOSCOPE EARPHONES   *Output No. 1*

STET 1	For models using Grundig jack plugs.
STET 3	For models using Grundig 3-pin plugs.
Type 207 (SE3)	Single earclip phone.
STET	Stethoscope adaptor for SE3.
Type 340 (SE6)	for monitoring and private listening.
Type 211	High quality stereophonic, with J21 plugs, for Grundig stereo recorders.
Type 220	High quality stereophonic with J21 plugs for hi-fi amplifiers.

REMOTE CONTROLS

Special Socket

RCH1	Hand operated, giving Stop/Start.	} Jack plug.
RCF1	Foot operated, giving Stop/Start.	
RCF2	Universal foot control, giving Stop/Start.	
RCF22	Universal foot control, giving Stop/Start/Backspace.	
RCF4	Universal foot control, giving Stop/Start.	
RCF30	Foot operated, giving Stop/Start. To be used in conjunction with Solenoid F40	> 5-pin plug.
RCF44	Universal foot control, giving Stop/Start/Backspace.	
RCF55	Universal foot control, giving Stop/Start/Backspace.	

MICROPHONE EXTENSION LEADS

MEC 5, -10, -15 — Extend Grundig microphone leads to 5, 10 and 15 yards respectively for 3-pin plugs.

MC 5, -10, -15 — Extend Grundig microphone leads to 5, 10 and 15 yards respectively for jack plugs.

DMC 6, -12, -18 — Extend Grundig Low impedance microphone lead to 6, 12 and 18 yards respectively.

TELEPHONE ADAPTORS

Diode input socket.

TA1 — For models using Grundig jack plugs.

TA3 — For models using Grundig 3-pin plugs.

SPARE LEADS

SL1 — Twin screened cable for all input and output jack socket connexions. Fitted jack plug and 2 wander plugs.

SL2 — Twin screened cable. Replaces SL1.

SL3 — Twin screened cable for all input and low impedance output connexions. Fitted 3-pin plug and 2 wander plugs.

SL33 — Twin screened cable for input or high or low impedance output connexions. Fitted 3-pin plug and 3 wander plugs, code: RED/BLACK—High Impedance, YELLOW/BLACK—Low Impedance.

SL233 (237) — Diode lead. For connecting to Grundig radios for recording and playback. Fitted 2 3-pin plugs.

SL3R	As SL3 with ½ M series resistor.
SL3X	Same as SL3 but 14 ft. long. For connecting Grundig Stereo models TK50 and TK55 to second channel reproducer. Also connects to extension speaker for Monaural reproduction.
SL33S	Twin screened cable fitted 3-pin plug and 3 free ends. For all input and output connexions to TK60. Or High (red-black) or Low (yellow-black) impedance output connexions.
SL233S	Twin screened cable fitted 2, 3-pin plugs. For all input and output connexions from TM60 deck to Grundig stereograms.
SL154	Twin screened cable fitted 5-pin plug and 4 free ends. For all input (red-common) and output (yellow-common) connexions to Cub.
SL142R	Screened cable fitted 4-pin plug and 2 free ends for recording powerful signals with Cub. Plug incorporates attenuating network.
SL144	Twin screened cable fitted 4-pin plug and 4 free ends. TK1 version of SL154.
SL132R	Screened cable fitted 3-pin plug and 2 free ends. TK1 version of SL142R.
SL155	Twin screened cable fitted 5-pin plug and 5 free ends for all input and high impedance output connexions to TK46.
SL255 (242)	Twin screened cable fitted 2, 5-pin plugs. For connecting TK46 to a Grundig stereo radio or gram fitted with a Diode Socket.
SL276	Twin screened cable fitted one 3-pin plug and 2, 3-pin sockets. For connecting separate earphones to each stereo channel of TM45 chassis.
SL277	Twin screened cable fitted one 3-pin plug and one 5-pin plug. For connecting Mixer Unit 608 to TM45 chassis.
SL30	Identical with SL33 but has no wander plugs.
SL32R	Recording lead as SL3 with ½ meg. resistor in series and 22K ohm across 1 and 2.
SL371R	Twin screened cable fitted two 3-pin plugs and incorporating resistance network in recording cable. For connecting Grundig tape recorder to Diode socket of radio for 2-way Record/Playback operation.

CHANNEL REPRODUCER CRI

Self-contained speaker/amplifier unit for high quality reproduction.

PLUGS

J1	6 mm. co-axial jack plug for all models with jack-type input and output sockets.
J11	Replaces J1.
J2	5-blade plug for remote controls.
J3	Grundig 3-pin shrouded plug for all input and output sockets of models with 3-pin sockets.

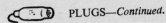 PLUGS—*Continued.*

J5	5-pin; for Remote Control lead.
J6	Earth plug for connecting early models to earth.
J12	Grundig Cub external battery connexion.
J14	All input and output connexions to 5-pin sockets.
SJS1	Screened socket designed to take both Grundig 6 mm. and British standard $\frac{1}{4}$ in. dia. jack plugs.
SJS3	Screened socket designed to take Grundig 3-pin plugs.
J17	Model TK1 and TK6 connexion to external battery.
J21	Output connexions to extension speaker or earphones, models TK46, TK18, TK18L, TK125, TK140, TK340 and TS340.
LA53	Adaptor for connecting 3-pin plug to Cub input/output socket.

DISTRIBUTOR SPEAKER

Twin cone high frequency speaker for improving high note response of model TK830/3D.

MIXER UNITS

GMU1	Mixes 4 inputs—2 condenser microphones and 2 sound channels. 4 control knobs for independent adjustment of inputs. 3 pre-set level controls.
GMU2	Mixes 4 inputs—2 Grundig dynamic microphones and 2 sound channels. 4 control knobs for independent adjustment of inputs, 2 pre-set level controls.
GMU3	Electronic mixer for mixing 4 inputs—2 (or 2 pairs of) Grundig condenser microphones and 1 high impedance dynamic or ribbon microphone, and 1 sound channel, giving 34 and 24 dB lift on dynamic and condenser sockets respectively and 12 dB attenuation on sound channel. Suitable for all Recorders.
Stereo Mixer 608	Transistorised stereo and monaural mixer for mixing stereo microphone and stereo radio or radiogram.
422	4-input, transistorised stero mixer.

MAINS PACK

MPP1	Converts Cub to mains operation.
MPP2	Converts TK1 to mains operation.
MPP100	Converts C100 to mains operation.

CAR BATTERY UNITS

AA380	6-12-24v operation for TK6.
AA386	12v operation for TK6, TK6L and C100.

MONITOR AMPLIFIER

MA1 Supplementary plug-in amplifier for monitoring Tracks 1 or 2 when recording Tracks 3 or 4 on Grundig 4-track model TK24.

MA2 As above for TK174, TK23, TK23L, TK40, TK140 and TK400. (Not interchangeable with MA1).

SONO-DIA SLIDE CHANGE ADAPTOR

SD1 Enables tape recorder to be used for automatically operating slide changer of transparency projector.